# Lecture Notes in Mathematics

Edited by A. Dold and B. Eckmann

881



Robert Lutz
Michel Goze

# Nonstandard Analysis

## A Practical Guide with Applications

Springer-Verlag
Berlin Heidelberg New York 1981

**Authors**

Robert Lutz
Michel Goze
Institut des Sciences Exactes et Appliquées, Université de Haute Alsace
4, rue des Frères Lumière, 68093 Mulhouse - Cedex, France

AMS Subject Classifications (1980): 03 H xx, 34 E xx

ISBN 3-540-10879-3 Springer-Verlag Berlin Heidelberg New York
ISBN 0-387-10879-3 Springer-Verlag New York Heidelberg Berlin

CIP-Kurztitelaufnahme der Deutschen Bibliothek
Lutz, Robert:
Nonstandard analysis: a pract. guide with applications / Robert Lutz; Michel Goze.–
Berlin; Heidelberg; New York: Springer, 1981.
(Lecture notes in mathematics ; Vol. 881)
ISBN 3-540-10879-3 (Berlin, Heidelberg, New York)
ISBN 0-387-10879-3 (New York, Heidelberg, Berlin)
NE: Goze, Michel:; GT



2141/3140-543210

## Dialogue

On the University's premises,

- Are you familiar with the non-standard methods ?
- ???
- Non-standard Analysis, if you prefer
- I've heard about ... Infinitesimals and the like, that seem to be coming back into fashion. But why call it a method ?
- Because it's not so much a question of bringing infinitesimals into fashion, but rather of furnishing a new proving tool, a non-standard one for those unfamiliar with it.
- Do you intend to modify our standards of reasoning ? You would'nt be the first... But old math is good enough for me !
- Not at all ! The non-standard method introduces intermediate objects - infinitesimals, for instance - by means of a very simple trick of language, in order to simplify the proofs - mainly whenever asymptotic behaviours are concerned - and to make them close to heuristic approaches.
- But you add something ! It's no longer the same math ! Enlarging the frame so as to introduce everything you want ... no wonder that your reasoning becomes simpler ! But is it still valid ?
- Undoubtedly ! Non-standard mathematics is strictly equivalent to the standard one : every non classical reasoning about the usual mathematical objects is equivalent to a classical one.
- So, there is nothing new in your method ! Because if you have a non classical proof, there is also a classical one. I prefer to search for the latter. Your method is pointless : either you prove known results or you produce a new result that you could, after all, prove in another way. What's the gain ?
- Your are one of those who prefer to march up towards the source of the river to get accross, instead of using a bridge down-stream. Because that is what it amounts to : to use bridges in order to avoid circuitous paths.

Recall Leibniz : "On ne diffère du style d'Archimède que dans les expressions, qui sont plus directes dans notre méthode et plus conformes à l'art d'inventer...".

Replace "Archimède" by "Bourbaki" and everything is said.

- Well ! But I heard that your method is quite complicated ; a logician's affair with languages, models and all that stuff... It seems that Robinson's book begins with fifty pages on logic in order to justify this infinitesimal nonsense !
- Don't worry about those pages ! They are conceived for the specialist's sake, to make the foundations irrefutable ; but you, as a mathematician, you should read the sequel, with all its promising developments.
- I'll try ; but really, I'm not convinced that non-standard methods may seriously change the mathematical landscape. Apart from esthetical and historical aspect, where's the importance for non-specialists ?
- True, if you only use Non-Standard Analysis to get easier presentations of well-known classical topics. But the main interest is elsewhere, in the fascinating world of applied mathematics ! For instance, engineers have to master a lot of perturbation phenomena for which classical tools are rather hard to work out. Due to its new intermedia - like infinitesimals - and powerful principles, Non-Standard Analysis allows deep investigation of perturbations in quite a natural way...
- I'm ready to agree, if you provide some examples. Meanwhile, I wonder by which miracle new objects may be introduced without further changes. Most likely these objects were already present, but they couldn't be talked about.
- Right ! Usually, to introduce an object, one begins with a definition what is non-standard is to introduce undefined objects, together with suitable restrictions.

For instance, the statement "there exists an element $\omega$ in the set of integers, larger than any integer which may be constructed by means of "adding a stroke", defies well established beliefs. For everybody, N is the collection I, II, III, etc...

However, we may consistently use such an undefined $\omega$, subject just to the conditions $\omega > \text{I}$, $\omega > \text{II}$, $\omega > \text{III}$, etc... ; indeed, in a proof involving $\omega$, one only uses a finite number of the above conditions ; hence the argument is still valid if one replaces $\omega$ by some "genuine" large enough integer.

Thus, $\omega$ doesn't disturb arithmetic and it deserves to be called infinitely large, as do $\omega$ - I, $\omega$ - II, $\omega$ - III or $\omega$ + I, $\omega$ + II, $\omega$ + III, ...

Similarly, the Non-Standard method introduces the undefined predicate "standard" with some restrictions in order to use it without betraying classical mathematics.

- Now I'm really lost ! What's this swindle about the set of integers ? Never heard such a nonsense !!

In the shadow of a grove,

Lutz : Here's a gentleman who is ready for our yellow booklet !
Goze : Indeed, but now he is angry about this story with $\omega$ ...
Lutz : Don't worry, it's only the initial shock necessary to get a non-standard mind ! he'll survive ...
Goze : Hey ! look who is coming there ! It's Georges with the catch-word ...

Exeunt omnes.

The authors wish to thank all those who have had a beneficent influence on the shape of this book, in particular

- all those non-standard minded people at Strasbourg - Oran - Mulhouse whose works are basic references to section IV ;
- our friends Dr. Tewfik SARI whose everyday collaboration was inestimable and Dr. Wilfried REYES whose linguistic improvements made our "formal" english text close to an actual one ;
- Prof. W.T. VAN EST and Prof. E.M. de JAGER whose kind invitation to Amsterdam and repeated encouragement strongly stimulated our developments on asymptotics ;
- Miss Huguette HAUSHALTER who typed the text with care and high efficiency

Of course, it is not possible to express our gratitude to Prof. Georges REEB in a few words. A few years ago, as he claimed in every corridor that Non-Standard Analysis was "something new" that should be a worthy tool in perturbation problems ; we believed it and thus had to write a book on it...

That he is so completely in the right and so interested in developing non-standard methods is not surprising : his intuition is legendary and he has the odd belief that mathematical research should be a pleasant adventure !

Mulhouse, Leimbach, Rammersmatt,

March 1981

One aim of this work is to stimulate a large discussion within the mathematical community about the efficiency of non-standard Analysis as a tool for mathematicians.

Therefore the authors invite the readers to send them their remarks (negative or positive...) both on the subject itself and on the topics involved in the present book.

R. Lutz and M. Goze

C O N T E N T S

## IV. NON STANDARD ANALYSIS AS A TOOL IN PERTURBATION PROBLEMS

A Christianne et Ginette

et à nos enfants

Emmanuelle

Joelle

Yannik

Christine

Emmanuelle

"The widely held belief that one cannot
get something for nothing is a superstition"
E. NELSON, BAMS 83 (1977), p. 1184.

Fîhr nix, ebb's ; diss gibbt's.

# FOREWORD

The dream of an infinitesimal calculus worthy of the name, that is to say in which dx and dy are infinitesimal numbers, $\int_a^b f(x)\,dx$ is a genuine sum of such numbers, limits are attained (or almost), formulae of type

$$\varphi(x) = \int_0^x f(\xi, \varphi(\xi - \tau))\, d\xi ,$$

with $\tau$ infinitely small furnish the solution of $y' = f(x, y)$ that satisfies $\varphi(0) = 0$ (*), has always been dreamed by mathematicians and such a dream deserves perhaps an epistemological inquiry.

Some other dreams, lesser maybe if compared with the achievements of calculus, have haunted the mathematician's imagination and wishful thought :
it is the idea of a world where integers can be classified as "large", "small" or even "indeterminate" without the loss of consistent reasoning, satisfy the induction principle and where the successors of small integers would remain small (**) ; a world where concrete collections, fuzzy perhaps but anyhow not finite, could be gathered in a single finite set ; a world where continuous functions would be approximated almost perfectly by polynomials of a fixed degree. In such a world, the finite realms could be explored either through the telescope or through the magnifying glass in order to gather entirely new pictures. Within such a world, the criteria of rigor set forth by Weierstrass and Göttingen, interpreted in a two-fold sense would allow for phantasy and metaphor.

This foreword is an opportunity to set forth the following remarks :

---

(*) This list may be extended: where $\frac{1}{\sqrt{2\pi\sigma}} \exp(-\frac{x^2}{2\sigma^2})$ for $\sigma > 0$ and $\sigma$ infinitesimal would serve as a Dirac function, where the teratology of the solutions of $y' = f(x, y)$, with continuous, non Lipschitz $f$ could be viewed as the regularity, seen through some appropriate glass, of the case in which f is analytic...

a) The outstanding work of A. Robinson on Non-Standard Analysis provides an astonishingly easy answer to this wishful dream. His book is still remarkable for the examples chosen in various fields of Mathematics or theoretical physics ; these examples were scarcely noticed, for questions of foundations seemed to be much more important.

b) The present work has a peculiar flavour among the books on Non-Standard Analysis published. R. Lutz and M. Goze's developments, centered around the idea of perturbation, singular perturbation and deformation, may show how the situation has evolued ever since ; although their book is research oriented, readers should acquire a good working knowledge of Non-Standard Analysis.

c) Developments arising from Non-Standard Analysis won't be utterly surprising to those willing to subscribe to the following simplified version of Brouwer, Skolem and Gödel :

Concrete sets in formalized mathematics do not cope with those provided by formalization.

d) Although the various pieces entering in c) are known since the twenties, mathematicians were not convinced that they could gather valuable results on the basis of c).

If we return to the quotation from Nelson that we have chosen as a heading, one may wonder why man disregards the use of such free gifts, whose very existence is undeniable.

Georges H. REEB

March 1981

---

(**) In chemistry for instance, ratios $p/q$ with "small" integers $p$, $q$ used to be considered. A distinguished mathematician gives a pleasant example :
"There should be a finite chain linking some monkey to Darwin, respecting the rules : a monkey's son is a monkey, the father of a man is a man."
Other examples could be found in the domain proper to "linguistics".

## READING GUIDE

This book is intended to enable the reader to use Non Standard Analysis by himself without fear, at any level of mathematical practice, from undergraduate analysis to important research areas.

It is divided into four sections with complementary purposes.

In Section I, the concept of enlargement with transfer and idealisation properties is introduced gradually and used to prove some statements on elementary calculus. To avoid a formal non motivated definition, we surround this concept with a progressive "order of procedure" as a hand rail.

In Section II, after a quick survey of set theory and some disturbing remarks about the gap between the potential collection of "natural" integers and the formal set $\mathbb{N}$ , enlargements are justified as by-products of the axiom of choice.

This study leads to a description of internal set theory (I.S.T.), an axiomatic approach to Non-Standard Analysis, which provides our game with pleasant rules. The existence of enlargements is closely related to the consistency proof of I.S.T., for it provides models of this theory.

Both approaches - with enlargements or within I.S.T. - are equivalent as regards mathematical praxis, and section III begins with a comparison of both working on general topology. For practical reasons, we go on using I.S.T., after a very small improvement to allow external sets in the discourse.

The remaining lessons of section III are devoted to a non-standard treatment of some important chapters of topology and differential calculus.
At this point, the reader should be able to use N.S.A. in whatever areas of mathematical research, in which it may be efficient.

Section IV is intended to give some recent examples of such attempts about various perturbation problems in algebra and differential equations ; here N.S.A. appears as an important tool in applied mathematics, according to the original aim of Abraham Robinson. No familiarity with classical perturbation theories is required, but some insight into the literature would make comparisons possible.

Section IV begins with a check-list of what is necessary to work in it without any knowledge of I, II, III. Of course, if you are not in a hurry, it is better to start with reading lesson 0, Section I.

The style of the book is rather non linear. Every lesson - some readers may wonder at this old-fashioned word, but we like its flavour - is focused on an essential information ; various comments, including proofs if necessary, invite the reader to ponder over this information in the light of his growing non-standard knowledge.

Some exercises sprinkle the text and there are topics to be developed.

A small glossary collects the terms which play some part in the book.

---

One aim of this book is to stimulate a large debate among mathematicians about the use of non-standard Analysis in the current research. Therefore we heartily invite the readers to send us their reactions (even bad ones, of course...) or tell us their own experiences with N.S.A..

PART I : ELEMENTARY PRACTICE OF NON STANDARD ANALYSIS

# Lesson 0

## (quite classic)

## INFINITESIMALS

Let $K$ be a commutative totally ordered non archimedean field, containing $R$ as an ordered subfield. To $K$ we associate :

i) the ring $F$ of finite elements ; $R$ is a subring of $F$ .

ii) the set $I$ of infinitesimals ; $I$ is an ideal of $F$ and $R \cap I = \{0\}$ .

iii) the set of infinitely large elements : that is $K-F$ .

iv) the equivalence relation $\sim$ (read "infinitely near") on $K$ .

v) a natural injection of ordered rings $\Phi : R \longrightarrow F/I$ .

THEOREM. $\Phi$ is an isomorphism, that is $F = R \oplus I$ , or finite elements are infinitely near to elements of $R$ .

Comments. 1) In a non archimedean field "the tortoise cannot overtake any hare", because there is an $\omega$ larger than every integer. Such odd fields exist ; the smallest one which contains $R$ is $R(X)$ , the field of rational fractions endowed with the degree relation.

2) As $R$ is archimedean, $\omega$ is not only larger than every integer, but also larger than every element of $R$ . Call infinitely large those elements of $K$ whose absolute value has this property. All other elements are called finite : each of them is bounded by a real number. The inverses of infinitely large elements together with $0$ are called infinitesimals. Sentences i) and ii) summarize the computing rules on infinitesimals ; they imply "(infinitely large) $\times$ (non infinitesimal) = (infinitely large)". Of course, a product of type "(infinitely large) $\times$ (infinitesimal)" may take any value.

3) Define $\sim$ by " $x \sim y$ if and only if $x - y \in F$ ". This equivalence relation is compatible with addition, but not with multiplication, for $I$ is not an ideal of $K$ (only of $F$ ).

4) Properties i) to v) are obvious. Let us prove the theorem :

If $a$ is a finite element of $K$ , the set $E = \{x \in \mathbb{R} , x \leq a\}$ is bounded from above ; its least upper bound ( $\mathbb{R}$ is complete !) is infinitely near to $a$ .

Thus we have an "infinitesimal calculus", but what can we do with it ?

---

## Lesson 1

(with a slight non standard flavour)

### LIMITS

THEOREM. " $u_n \longrightarrow 0$ as $n \longrightarrow \infty$ " is equivalent to "for every infinitely large $n$ , $u_n$ is infinitesimal".

Comments. 1) Within $\mathbb{R}$ , the second part of this sentence obviously has no sense : there are no infinitesimals but $0$ , if we agree with ABEL's definition. For this reason, the infinitesimal point of view in analysis did not survive, except as a figure of speech, after CAUCHY and WEIERSTRASS replaced it - with regret - by the wellknown " $\varepsilon - \delta$ " concept of limit.

2) In the frame of Lesson 0, the sentence makes sense, but, if we consider $u_n$ as a real sequence, we have first to select a set $\widetilde{\mathbb{N}}$ of "generalized integers" in the non euclidean field $K$ , which contains $\mathbb{N}$ and also infinitely large elements ; then we have to extend the mapping $u : \mathbb{N} \longrightarrow \mathbb{R}$ to a mapping $\tilde{u} : \widetilde{\mathbb{N}} \longrightarrow K$ (which in the sentence above is improperly named $u$ ).

3) Take for instance $\tilde{u}_n = u_n$ if $n \in \mathbb{N}$ and $\tilde{u}_n = 0$ if $n \in \tilde{\mathbb{N}}$ is infinitely large. Then it follows from the theorem that every real sequence has limit $0$. This would considerably simplify real analysis, wouldn't it ?

4) Thus, to give our theorem some chance to be true, we have to relate closely the properties of $\tilde{u}$ with those of $u$.

Let us outline a proof based on such a demand. For the direct part, consider $\varepsilon > 0$, $\varepsilon \in R$. Then there is an $n_o \in \mathbb{N}$ such that the map $u$ has the property " $\forall n > n_o$, $n \in \mathbb{N} \Longrightarrow |u_n| < \varepsilon$" . Suppose that the related property is true for $\tilde{u}$, that is " $\forall n > n_o$, $n \in \tilde{\mathbb{N}}$, $|\tilde{u}_n| < \varepsilon$ ". Thus, for every infinitely large $n$, we have $|\tilde{u}_n| < \varepsilon$, which implies $|\tilde{u}_n| \sim 0$.

Conversely, suppose $|\tilde{u}_n| \sim 0$ for every infinitely large $n \in \tilde{\mathbb{N}}$. Then the property " $\forall n > n_o$, $n \in \tilde{\mathbb{N}} \Longrightarrow |\tilde{u}_n| < \varepsilon$ " is true for every $\varepsilon > 0$, $\varepsilon \in R$ and $n_o$ infinitely large. Hence, for every fixed $\varepsilon$, the statement " $\exists n_o$, $\forall n > n_o$, $n \in \tilde{\mathbb{N}} \Longrightarrow |\tilde{u}_n| < \varepsilon$ " is true. If the relation between $u$ and $\tilde{u}$ is sufficient, we may hope that it remains true if we replace $(\tilde{u}, \tilde{\mathbb{N}})$ by $(u, \mathbb{N})$. This would end the proof.

Note the correct formulation of the theorem :

THEOREM. " $u_n \longrightarrow 0$ as $n \longrightarrow \infty$ ", <u>where</u> $u_n$ <u>is a real sequence</u>, <u>is equivalent to "for every infinitely large</u> $n \in \tilde{\mathbb{N}}$, $\tilde{u}_n$ <u>is infinitesimal in</u> $K$ ".

5) Consider for instance the sequence $u_n = \frac{1}{n}$. We have no information about $\tilde{u}_n$ for infinitely large $n$. But we certainly would ask for an $\tilde{u}$ which extends the property " $nu_n = 1$ ", that is to take $\tilde{u}_n = \frac{1}{n}$. Then our characterization of limits leads to $\frac{1}{n} \longrightarrow 0$ (of course, this is a rather complicated way to prove it...).

Our business is to find $K$, $\tilde{\mathbb{N}}$, and an extension rule $u \longrightarrow \tilde{u}$ so that all this works. Fortunately, we have an answer (within our classical mathematics !). Its key word is <u>enlargements</u> and it was A. ROBINSON's idea to build on this concept a new procedure in Analysis. Little by little, we shall endeavour to instruct the reader on the use of enlargements. A precise definition is given further on (sect. II).

Order of procedure. Article 1. The following properties are satisfied in an enlargement $^*\mathbb{R}$ of $\mathbb{R}$ :

- every part $A \subset \mathbb{R}$ has a natural extension $^*A \supset A$ .
- every mapping $u : A \longrightarrow \mathbb{R}$ has a natural extension $^*u : {}^*A \longrightarrow {}^*\mathbb{R}$.
- the operations $+$ , $\times$ and the order relation $<$ have natural extensions $^*+$ , $^*\times$ , $^*<$ , working on $^*\mathbb{R}$ .
- more generally, every binary, ternary, ... relation on $\mathbb{R}$ has a natural extension to $^*\mathbb{R}$ .

Article 2. In these extensions, every property which can be stated by means of the above, together with logical connectors and with (eventually quantified) variables running only over the elements, is true on $\mathbb{R}$ if and only if it is true on $^*\mathbb{R}$ provided the ingredients are replaced by their extensions. The variables shall not be starred.

As an example, consider :

$$\forall x \quad (((x \geq 0) \Longrightarrow (|x| = x)) \wedge ((x \leq 0) \Longrightarrow (|x| = -x))) .$$

This is true on $\mathbb{R}$ and translated to $^*\mathbb{R}$ gives :

$$\forall x \quad (((x \ {}^*\!\!\geq 0) \Longrightarrow ({}^*|x| \ {}^*\!\!= x)) \wedge ((x \ {}^*\!\!\leq 0) \Longrightarrow ({}^*|x| \ {}^*\!\!= {}^*\!- x))) .$$

Article 2 bis. One can forget about the stars if no confusion is likely to arise.

Article 3. $^*\mathbb{R}$ is a totally ordered non archimedean extension of $\mathbb{R}$ .

Exercises. 1) Use articles 1 and 2 to prove that $^*\mathbb{R}$ is a commutative totally ordered field.

2) Try to prove article 3.

3) Give up, but not too easily, and proceed to article 5 in lesson 6.

# Lesson 2

(entirely non standard)

## CONTINUITY

THEOREM 1. <u>A mapping</u> $f : A \subset R \longrightarrow R$ <u>is continuous at a point</u> $x_o$ <u>if and only if for every infinitesimal</u> $\eta$, ${}^*f(x_o + \eta) \sim f(x_o)$ .

THEOREM 2. <u>Let</u> $f : [0, 1] \longrightarrow R$ <u>be a continuous mapping such that</u> $f(0) < 0$ <u>and</u> $f(1) > 0$ . <u>Then there is a point</u> $c$ <u>in</u> $]0, 1[$ <u>such that</u> $f(c) = 0$ .

<u>Comments</u>. 0) Recall that $x \sim y$ means that $x - y$ is infinitesimal.

1) If there is no danger of confusion, we may forget the stars. For instance, we write $+$ , $x$ , $<$ , $-$ , $|\ |$ , instead of ${}^*+$ , ${}^*x$ , ...

2) Let us use theorem 1 to prove the continuity of $f(x) = ax$ , with $a \in R$ . By transfer, ${}^*f$ is again the multiplication by a working on ${}^*R$ . Thus, if $\eta \sim 0$ , we have ${}^*f(x_o + \eta) = ax_o + a\eta \sim ax_o = f(x_o)$ . According to theorem 1, $f$ is continuous.

3) If $x_o \in R$ , we call <u>halo</u> ("monad" in [R] ) of $x_o$ , the set $h(x_o) = \{x \in {}^*R , x \sim x_o\}$ . By use of the halo concept, continuity at $x_o$ is equivalent to ${}^*f(h(x_o)) \subset h({}^*f(x_o))$ .

Notice the "covariant" form of this definition and compare it with the usual one.

4) The mapping $f : A \longrightarrow R$ is transfered into a mapping ${}^*f : {}^*A \longrightarrow {}^*R$ . For instance, if $A = [a, b] = \{x \in R / a \leq x \leq b\}$ , we get, by transfer, using article 2, ${}^*[a, b] = \{x \in {}^*R / a \leq x \leq b\}$, i.e. an interval of ${}^*R$ .

Proof of theorem 1. 1) Suppose that $f : A \longrightarrow R$ is continuous for $x = x_0$ or, in plain english :

$$\forall \varepsilon , [(\varepsilon > 0) \Rightarrow \exists \eta , \forall x ( |x - x_0| < \eta \Rightarrow |f(x) - f(x_0)| < \varepsilon)] .$$

Let $\varepsilon$ be fixed in $R$ ; transfer the property

$$\forall x , |x - x_0| < \varepsilon \Rightarrow |f(x) - f(x_0)| < \varepsilon .$$

Then, if $|x - x_0|$ is infinitesimal, we have $|x - x_0| < \eta$ , hence $|{}^*f(x) - {}^*f(x_0)| < \varepsilon$ ; as $\varepsilon$ is arbitrary, we conclude that ${}^*f(x) \sim {}^*f(x_0)$ .

2) Conversely, let $\eta$ be an infinitesimal. Then for every $\varepsilon > 0$ , $\varepsilon \in R$ , we have :

$$\forall x , |x - x_0| < \eta \Rightarrow |{}^*f(x) - {}^*f(x_0)| < \varepsilon .$$

Thus the property

$$\exists \eta \text{ such that } |x - x_0| < \eta \Rightarrow |{}^*f(x) - {}^*f(x_0)| < \varepsilon$$

is true in ${}^*R$ .

By transfer, the property

$$\exists \eta \text{ such that } |x - x_0| < \eta \Rightarrow |f(x) - f(x_0)| < \varepsilon$$

is true in $R$ .

Proof of theorem 2. We fix an infinitely large $\omega \in {}^*\mathbb{N}$ and consider the set $A = \{n \in {}^*\mathbb{N} , n \leq \omega \text{ and } {}^*f(\frac{n}{\omega}) \geq 0\}$ . Then $\omega \in A$ and $A$ has a smallest element $n_0$ and, as $0 < \frac{n_0}{\omega} \leq 1$ , there is a $c \in R$ such that $\frac{n_0}{\omega} \sim c$ . We have $0 \leq c \leq 1$ , $\frac{n_0 - 1}{\omega} \sim \frac{n_0}{\omega} \sim c$ ; hence, by continuity, ${}^*f(\frac{n_0 - 1}{\omega}) \sim f(c) \sim {}^*f(\frac{n_0}{\omega})$ which proves that $f(c) = 0$ , for both extremes have opposite sign.

Remarks. 1) There is a cheat in the above proof : in lesson 8, we shall see that there are subsets of ${}^*\mathbb{N}$ without a smallest element ! A could be one of those ! But the cheat is temporary : A being defined by a "nice" property is an internal part (see lesson 9, section I) ; by transfer, internal parts inherit the properties of the parts of $\mathbb{N}$ .

2) Theorem 1 is non standard. Theorem 2 is standard, but it is proved by non standard methods.

IMPROVED ORDER OF PROCEDURE.

Article 2 ter. From now on, instead of "putting stars", we shall say transfer. Its use in the proof of theorem 1 is typical.

- A transfered property is true in $^*R$ if and only if it is true in $R$ .

- Recall once more that the variables are not starred.

- Each element of $R$ will be called standard.

Article 4. Let $x_o$ be in $R$ . The set $h(x_o) = \{x \in {}^*R \mid x \sim x_o\}$ is called halo of $x_o$ .

By means of a powerful magnifying glass, one can see in the halo of $x_o$ properties common to all the neighbourhoods of $x_o$ .

---

Lesson 3

UNIFORM CONTINUITY

THEOREM 1. A function $f$ is uniformly continuous on a subset $A$ of $R$ if and only if for every $x , y \in {}^*A$ , $x \sim y$ implies $^*f(x) \sim {}^*f(y)$ .

THEOREM 2. A continuous function on a closed interval $[a , b]$ is uniformly continuous on it.

Examples. 1) Consider the function $\sqrt{x}$ defined on $R^+$ . The square of the function transfered to $^*(R^+) = (^*R)^+$ is $x$ . Thus we call this function $\sqrt{x}$ . If $x \geq y$ , we have $0 \leq \sqrt{x} - \sqrt{y} \leq \sqrt{x-y}$ ; if $x \sim y$ , then $\sqrt{x} \sim \sqrt{y}$ and we infer from theorem 1 that $\sqrt{x}$ is uniformly continuous.

2) The function $x^2$ on $R$ is not uniformly continuous. Indeed for

an $\varepsilon \sim 0$ , consider $x = \frac{1}{\varepsilon} + \varepsilon$ , $y = \frac{1}{\varepsilon}$ . Then $x^2 - y^2 = 2 + \varepsilon^2$ which is not infinitesimal.

Comments. 1) Theorem 1 points out the difference between continuity and uniform continuity. In the former case, $x$ is a standard point of $A$ and $y$ a point of $^*A$ ; in the later case, $x$ and $y$ run freely through $^*A$ . In example 2, the pair $(\frac{1}{\varepsilon} + \varepsilon , \frac{1}{\varepsilon})$ is not in $A \times {}^*A$ and continuity remains possible.

2) Proof of theorem 1. Consider $\varepsilon > 0$ , $\varepsilon \in R$ . From uniform continuity, we get a $\delta > 0$ , $\delta \in R$ with the property

$$\forall x , \forall y , ((x \in A , y \in A \text{ and } |x - y| < \delta) \Longrightarrow (|f(x) - f(y)| < \varepsilon)) .$$

By transfer we have, for $^*f : {}^*A \longrightarrow {}^*R$ :

$$\forall x , \forall y , ((x \in {}^*A , y \in {}^*A \text{ and } |x - y| < \delta) \Longrightarrow (|{}^*f(x) - {}^*f(y)| < \varepsilon)) .$$

Thus $x \sim y$ implies $|{}^*f(x) - {}^*f(y)| < \varepsilon$ for every $\varepsilon \in R$ , $\varepsilon > 0$ , that is $^*f(x) \sim {}^*f(y)$ .

Conversely, consider again $\varepsilon > 0$ , $\varepsilon \in R$ . The property

$$\exists \delta \ \forall x \ \forall y \ [(x \in {}^*A \ ; \ y \in {}^*A \text{ and } |x - y| < \delta) \Longrightarrow (|{}^*f(x) - {}^*f(y)| < \varepsilon)]$$

is true in $^*R$ (take $\delta \sim 0$ ) ; this is the transfered form of the uniform continuity property for $f$ .

3) Proof of theorem 2. Consider $x$ and $y$ in $^*[a , b]$ with $x \sim y$ . As $a$ and $b$ are standard, $x$ is finite ; thus there is an $c \in [a , b]$ such that $y \sim x \sim c$ and, by continuity, $^*f(y) \sim {}^*f(c) \sim {}^*f(x)$ which prove uniformity.

4) Remark. The set $[a , b]$ is compact, and above we used the fact that every point in $^*[a , b]$ is infinitely near some point of $[a , b]$ . This property extends to general compact sets, as we shall see further (III, lesson 3).

# Lesson 4

## DERIVATIVES AND INTEGRALS

THEOREM 1. A function $f : \mathbb{R} \longrightarrow \mathbb{R}$ is differentiable at $x_o$ if and only if there is a standard number $f'(x_o)$ such that for every $h \sim 0$ $(h \neq 0)$,

$$\frac{{}^*f(x_o + h) - {}^*f(x_o)}{h} \sim f'(x_o) .$$

THEOREM 2. Let $f : [a, b] \longrightarrow \mathbb{R}$ be continuous. Then

$$\int_a^b f(x)\, dx \sim \sum_{j=1}^{j=n} {}^*f(x_j)(x_j - x_{j-1})$$

where $(x_j)_{1 \leq j \leq n}$ is an infinitely fine subdivision of ${}^*[a, b]$.

THEOREM 3. Let $F : [a, b] \longrightarrow \mathbb{R}$ have continuous derivative. Then

$$\int_a^b F'(x)\, dx = F(b) - F(a) .$$

Comments. 1) Theorem 1 is evident, if we recall the characterization of limits in lesson 2. However, we may formulate it without refering to $f'(x_o)$ : we demand that each number $\frac{{}^*f(x_o + h) - {}^*f(x_o)}{h}$ shall be finite for every $h \sim 0$, and that any two of such numbers shall be infinitely close to each other ; the derivative is their common shadow.

2) Theorem 1 realizes the old dream of a calculus in which the integral was considered as an infinite sum of infinitely thin rectangles.

3) Proof of theorem 2 : Let $(u_i^n)_{1 \leq i \leq n}$ be a subdivision of $[a, b]$, whose maximal step has limit 0 as $n \rightarrow \infty$. We know that

$$\int_a^b f(x)\, dx = \lim_{n \to \infty} \sum_{i=1}^{n} f(u_i^n)(u_i^n - u_{i-1}^n) .$$

After transfer of the sequence and of $f$, we get

$$\int_a^b f(x)\, dx \sim \sum_{i=1}^{n} {}^*f(u_i^n)({}^*u_i^n - {}^*u_{i-1}^n)$$

provided $n$ is infinitely large.

4) Although the former proof is correct, it doesn't prove theorem 2 because we are not for the time being sure to obtain every infinitely fine subdivision of ${}^*[a\,,b]$ by translating a sequence $u_i^n$ . For instance, the proof works for the subdivision $\{a + \frac{i(b-a)}{\omega}\}_{a\le i\le\omega}$ , with $\omega$ infinitely large in ${}^*\mathbb{N}$ , which is obtained from the sequence $u_i^n = a + \frac{i(b-a)}{n}$ . More generally, it works for every internal subdivision (see lesson 9 for internal).

5) Theorem 3 is the "fundamental theorem of calculus". Let us prove it using infinitesimals : use the subdivision $a + \frac{i(b-a)}{\omega} = x_i$ . Then

$$ {}^*F(x_i) - {}^*F(x_{i-1}) = (x_i - x_{i-1})\ {}^*(F')(x_i + (x_i - x_{i-1})\theta_i) $$

with $\theta_i \in {}^*]0\,,1[$ (Transfer the mean value formula). As $F'$ is continuous, we get (lesson 3) ${}^*F'(x_i + (x_i - x_{i-1})\theta_i) = {}^*F'(x_i) + \varepsilon_i$ with $\varepsilon_i \sim 0$ . Thus

$$ \sum_{i=1}^{n} {}^*F'(x_i)(x_i - x_{i-1}) = F(b) - F(a) - \sum_{i=1}^{n} \varepsilon_i \frac{b-a}{n} $$

and

$$ \int_a^b F'(x)\,dx \sim F(b) - F(a) \quad \text{for} \quad \sum_{i=1}^{n} \varepsilon_i \frac{b-a}{n} \sim 0 \ ; $$

but two infinitely near standard numbers are equal.

Order of Procedure.

Article 4bis. The shadow of a finite element $x$ of ${}^*\mathbb{R}$ is the only standard real number which is infinitely close to $x$ . It will be written $^\circ x$ .

Meditation on continuity. The function $\sin\alpha x : {}^*\mathbb{R} \longrightarrow {}^*\mathbb{R}$ , where $\alpha$ is infinitely large in ${}^*\mathbb{R}$ has the following property :

$$ \forall\, \varepsilon > 0\ ,\ \ \varepsilon \in {}^*\mathbb{R}\ ,\ \ \exists\, \eta \in {}^*\mathbb{R}\ ,\ \ |x - x_0| < \eta \Longrightarrow |\sin\alpha x - \sin\alpha x_0| < \varepsilon $$

where $x_0 \in \mathbb{R}$ (transfer the properties of $\sin uv$ ). We say that $\sin\alpha x$ is $*$-continuous at $x_0$ .

But if we forget the stars, it's no longer true ! Now take $\alpha$ finite. Then both properties are satisfied. We say that the function is also

S-continuous at $x_o$ .

Exercises. 1) Find an S-continuous function which is not $*$-continuous.

2) Show that a function $f : R \longrightarrow R$ is continuous at $x_o$ iff ${}^*f$ is S-continuous (recall lesson 2) at $x_o$ .

---

Lesson 5

DIFFERENTIABILITY

Consider a mapping $f : R^p \longrightarrow R^q$ . For every point $x \in {}^*R^p$ , define $\Delta f_x : {}^*R^p \longrightarrow {}^*R^q$ by $\Delta f_x(h) = {}^*f(x+h) - {}^*f(x)$ .

PROPOSITION 1. $f$ is continuous at $a \in R^p$ if and only if for every $h \sim 0$ , $\Delta f_a(h) \sim 0$ .

PROPOSITION 2. $f$ is differentiable at $a \in R^p$ if and only if $\Delta f_a$ is approximately linear (briefly A.L.).

PROPOSITION 3. $f$ is continuously differentiable on $R^p$ if and only if

i) for every $x \in {}^*R^p$ , $\Delta f_x$ is A.L.

ii) if $a \in R^p$ and $x \sim a$ , then $\frac{1}{\|h\|}(\Delta f_x(h) - \Delta f_a(h)) \sim 0$ for every $h \sim 0$ , $h \neq 0$ .

PROPOSITION 4. $f$ is of class $C^\infty$ on $R^p$ if and only if for every $a \in R^p$ and every $r \in \mathbb{N}$ , $\Delta^r f_a$ is approximately $r$-linear ; the jet of $f$ at $a$ is given by the formula

$$D^r f_a(u_1, \dots, u_r) = {}^\circ\left(\frac{\Delta^r f_a(\varepsilon u_1, \dots, \varepsilon u_r)}{\varepsilon^r}\right) \text{ where } u_i \in R^p \text{ and}$$

$\varepsilon \sim 0$ , $\varepsilon \in {}^*R$ .

<u>Comments</u>. 0) Here ${}^*R^p$ is $({}^*R)^p$, where ${}^*R$ is an enlargement of $R$. Note that $p$ and $q$ are standard integers. Later, we shall consider enlargements whose basis is $R^p$ ; but in this lesson, it is quite useless, if we consider $({}^*R)^p$ as a vector space on ${}^*R$ provided with the "norm" $\|x\| = \sup|x_i|$ (which takes its values in ${}^*R$ ). An element $x = (x_1, \dots, x_p) \in {}^*R^p$ is called infinitesimal (write again $x \sim 0$ ) if and only if every $x_i$ is infinitesimal; this is equivalent to $\|x\| \sim 0$ .

Every element of finite norm (that is of finite components) has a standard shadow $°x = (°x_1, \dots, °x_p)$ . Of course $°(\lambda x + \mu y) = °\lambda °x + °\mu °y$ if $\lambda$ , $\mu$ , $\|x\|$ , $\|y\|$ are finite.

It is clear that we could complicate the lesson replacing $R^p$ by open subsets. But this is really not useful .

1) It is easy to generalize the non standard characterization of limits (lesson 1) : "$f(a+h) \longrightarrow \ell$ as $\|h\| \longrightarrow 0$" is equivalent to ${}^*f(a+h) \sim \ell$ for every $h \sim 0$ ( $a$ is standard). Proposition 1 is an immediate consequence.

2) It is clear that a continuous mapping $f : R^p \longrightarrow R^q$ is continuous in each variable, the converse being false. Indeed, for a 2 variable function, consider

$$\begin{aligned} {}^*f(x_0+\varepsilon, y_0+\eta) - f(x_0, y_0) &= {}^*f(x_0+\varepsilon, y_0+\eta) - {}^*f(x_0+\varepsilon, y_0) \\ &\quad + {}^*f(x_0+\varepsilon, y_0) - f(x_0, y_0) \end{aligned}$$

where $(x_0, y_0)$ is standard and $\varepsilon \sim \eta \sim 0$ .

The partial continuity is not strong enough to make the first difference infinitesimal, for ${}^*f(x_0+\varepsilon, \cdot) : {}^*R \longrightarrow {}^*R$ is generally not transfered from a continuous function on $R$ .

Let us call "<u>neo-continuous at</u> $u \in {}^*R$ " a function $g$ such that for every $\eta \sim 0$ , $g(u+\eta) \sim g(u)$ .

Then $f$ is continuous at $(x_0, y_0)$ if and only if, for every

$\varepsilon \sim 0$ , $\eta \sim 0$ the partial functions $^*f(x_o + \varepsilon), \cdot)$ and $^*f(\cdot, y_o + \eta)$ are neo-continuous at $y_o$ (resp. $x_o$ ). For instance, the function $\frac{xy}{x+y}$ cannot be continuously extended to the origin for $^*f(\varepsilon, 0) - ^*f(\varepsilon, \varepsilon) = \frac{1}{2}$ .

3) The usual definition of differentiability assumes the existence of a linear mapping $L : R^p \longrightarrow R^q$ such that

$$\frac{1}{\|h\|}\left[^*f(a+h) - ^*f(a) - L(h)\right] \sim 0 \quad \text{for all} \quad h \sim 0 .$$

The corresponding intuitive concept is rather :

" $\Delta f_a(h)$ is <u>approximatively linear</u> for a small $h$ ".

Non standard analysis allows us to formulate such a notion as follows : a map $\varphi : {}^*R^p \longrightarrow {}^*R^q$ <u>is called A.L.</u> if it satisfies the following conditions :

i) <u>for every finite</u> $\lambda, \mu$ <u>in</u> $^*R$ <u>and</u> $h, k \sim 0$ <u>in</u> $^*R^p$ , <u>there is an</u> $\varepsilon \sim 0$ <u>and an</u> $\eta \sim 0$ <u>in</u> $^*R^q$ <u>such that</u>

$$\varphi(\lambda h + \mu k) - \lambda\,\varphi(h) - \mu\,\varphi(k) = |\lambda|\,\|h\|\varepsilon + |\mu|\,\|k\|\eta \;;$$

ii) $\frac{\varphi(h)}{\|h\|}$ <u>is finite for every</u> $h \sim 0$ , $h \neq 0$ .

In other words, (i) means that the "lack of linearity" belongs to the F-module generated by the numbers $|\lambda|\,\|h\|$ and $|\mu|\,\|k\|$ over the F module of infinitesimals in $^*R^q$ ; recall that F is the ring of the finite elements of R .

The link between the differentiability of $f$ at $a$ and the approximate linearity of $\Delta f_a$ lies from the following remark :

<u>If</u> $g$ <u>is A.L., then the mapping</u> $\hat{\varphi}$ <u>of</u> $R^q$ <u>defined by</u> $\hat{\varphi}(u) = {}^\circ\left(\frac{\varphi(\tau\omega)}{t}\right)$ , $\tau$ <u>being a fixed infinitesimal</u>, <u>is linear</u>, <u>and independent of</u> $\tau$ .

Just compute $\Delta f_a(u) - \widehat{\Delta f}_a(u)$ , $u \in R^p$ :

4) Clearly A.L. mappings are good behaved with respect to sums, products, composition. It is a pleasant exercise to deduce from this the cor-

responding rules for differentiability. Proposition 3 is proved by transfer from the mean value theorem and uses the equivalence relation $\frac{\varphi(h) - \varphi'(h)}{\|h\|} \sim 0$ for $h \sim 0$ between the A.L. mappings $g$ and $g'$.

Note that every equivalence class contains exactly one transfered linear mapping, which is $^*\varphi$. For instance, the continuity of $Df$ at $a$ means that for every $x \sim a$, $\Delta f(x)$ and $\Delta f(a)$ are in the same class.

5) Regarding higher-order differentiability, things are slightly more complicated due to the occurence of operators such as $D$, $\Delta$, $D^2$, $\Delta D$, $D\Delta$, ... and their higher-order analogues. Define the operator $\Delta^r$ by means of the recursive relation : $\Delta^{i+1} = \Delta(\Delta^i)$. Approximative r-linearity may be understood from the particular case $r = 2$ : we assume that :

i) $\varphi(\lambda_1 h_1 + \lambda_2 h_2 , \mu_1 k_1 + \mu_2 k_2) - \Sigma \lambda_i \mu_i \varphi(h_i , k_i)$

belongs to the F-modul generated by $\|\lambda_1 h_1\|$, $\|\lambda_2 h_2\|$, $\|\mu_1 k_1\|$, $\|\mu_2 k_2\|$.

ii) $\frac{\varphi(h , k)}{\|h\| \|k\|}$ is finite for $h \sim k \sim 0$.

The property is well behaved as regards the algebraic operations, in particular tensor products. The equivalence relation

$$\frac{\varphi(h, k) - \varphi'(h, k)}{\|h\| \|k\|} \sim 0$$

for $h \sim k \sim 0$ is compatible with these operations and in every class, there is one transfered r-linear mapping defined by $\hat{\varphi}(u , v) = {}^\circ\left(\frac{\varphi(\tau u , \tau v)}{\tau^2}\right)$ with $\tau \sim 0$.

The core of differential calculus lies on the fact that $\Delta^2 f_a$, $\Delta D f_a$, $D\Delta f_a$, and $D^2 f_a$ are equivalent in this sense.

The jet of $f$ at $a$ being given by $D^r f_a$, it is also given by $\hat{\Delta} f_a$ ; this justifies the second part of proposition 4, which describes its extraction.

Once again, the proof of the first part is an immediate consequence of the transfered mean-value formula:

6) Thus, differential calculus appears as "the shadow" of the infinitesimal difference calculus ; such an opinion has been maintained since a long time ago ; without anyone however in position to formulate it clearly.

---

## Lesson 6

## SOME NOTIONS OF TOPOLOGY IN $\mathbb{R}$

THEOREM. " $\ell$ is an accumulation point of the sequence $u_n$ " corresponds to "in $^*\mathbb{N}$ , there exists an infinitely large $\omega$ such that $u_\omega \sim \ell$ ".

Application. In $\mathbb{R}$ , every bounded sequence has an accumulation point (Bolzano-Weierstrass theorem).

Proof of Bolzano-Weierstrass theorem. We have $\forall\, n \ (|u_n| < M)$ in $\mathbb{R}$ . This property transfers to $^*\mathbb{R}$ . For each infinitely large $\omega$ , $u_\omega$ is finite, and according to the above theorem, its standard part $°u_\omega$ is an accumulation point of the sequence.

Comments: 1) Recall that the shadow $°\ell$ of a finite element $\ell$ is its standard part. A part of $^*\mathbb{R}$ will have as shadow the set of shadows of its finite elements. Clearly $\ell \in h(°\ell)$ , and if $a \in \mathbb{R}$ , the shadow $°(h(a))$ of the halo of $\{a\}$ is equal to $\{a\}$ .

If $f : A \subset {}^*\mathbb{R} \longrightarrow {}^*\mathbb{R}$ is such that all the elements of $f(A)$ are finite, we obtain a map $°f : °A \longrightarrow \mathbb{R}$ , the shadow of $f$ , defined by $°f(x) = °(f(x))$ .

Notice that a finite element is standard if and only if it is equal to its shadow. The old saying "one must not mistake the prey for the shadow" applies particularly well here : the properties of $f$ : continuity, differentiability, etc. (in a sense to be specified) cannot be transmitted to ${}^{o}f$ without resistance. We shall see later, in the theory of complex analytic functions how the shadow inherits the property of analycity : this example shows anew the strength of this property.

2) The above proof describes <u>all</u> the accumulation points of the sequence $u_n$ ; they are the shadows of its values for $n$ infinitely large. (From the standard point of view, this corresponds to all the limits of convergent subsequences.) Such a description becomes really enjoyable in the frame of functional spaces (see lesson 5, section III).

3) We must recall that ${}^{*}\mathbb{N}$ is here the enlargement of the subset $\mathbb{N}$ in $\mathbb{R}$ . Let us show now that <u>any finite element of</u> ${}^{*}\mathbb{N}$ <u>is standard</u>.

So let $n$ be a finite element of ${}^{*}\mathbb{N}$ . Then its shadow $\ell = {}^{o}n$ is an element of $\mathbb{R}$ , and there exists $n_o$ in $\mathbb{N}$ such that $n_o \leq \ell < n_o + \ell$ . Therefore we have $n_o - 1 < n < n_o + 1$ . But the only solution of $n_o - 1 < x < n_o + 1$ in $\mathbb{N}$ is $n_o$ .

By transfer, $n_o$ is also the unique solution in ${}^{*}\mathbb{N}$ : Hence $n = n_o$ .

4) Up to now the existence of infinitely large integers has been taken for granted, otherwise we would have come to naught. Don't worry, we'll soon come up to the point !

5) Even worse ! In the course of the proof, we came across the " $n$ infinitely great such that ... " . Here is the secret : consider the relation $\mathcal{R}$ with three entries $(\nu , \varepsilon , n)$ whose source is $\mathbb{N} \times \mathbb{R}^{+}$ .

" $\nu$ and $n$ are integers, $n > \nu$ and $u_n - \ell < \varepsilon$ ".

Then, for any <u>finite</u> family of pairs $(\nu_i , \varepsilon_j)$ in $\mathbb{N} \times \mathbb{R}^{+}$ , the number $\varepsilon = \inf \varepsilon_j$ is strictly positive, and there exists in $\mathbb{N}$ an integer $n > \nu = \sup \nu_i$ , verifying $\mathcal{R}(\nu , \varepsilon , n)$ , and so also $\mathcal{R}(\nu_i , \varepsilon_j , n)$ for any $i , j$ .

Under these conditions there exists in ${}^{*}\mathbb{N}$ (this is a virtue of ${}^{*}\mathbb{R}$

that must be taken for granted) an integer $\omega$ such that $^*\mathbb{R}(\nu, \varepsilon, \omega)$ for each $\nu \in \mathbb{N}$ and $\varepsilon \in \mathbb{R}^+$. Hence $^*u_\omega \sim \ell$.

6) Conversely, we invoke reversed transfer : $\nu$ and $\varepsilon$ being fixed (in $\mathbb{R}$), the property

" $\exists n$ ( $n$ being an integer, $n > \nu$ and $|u_n - \ell| < \varepsilon$ ) "

is true in $^*\mathbb{R}$ (take $n = \omega$), thus also in $\mathbb{R}$, which shows well that for each $\nu \in \mathbb{N}$ and $\varepsilon \in \mathbb{R}^+$, there is an $n > \nu$ in $\mathbb{N}$ such that $|u_n - \ell| < \varepsilon$.

7) If we admit this sort of argument, we can use it again to justify the existence of infinitely large elements of $^*\mathbb{R}$ or $^*\mathbb{N}$, and more generally of $^*A$, where $A$ is an unbounded subset of $\mathbb{R}$ : Indeed the binary relation $\mathcal{R}$ :

" $\rho(x)$ and $\rho(y)$ and $x < y$ "

of source $A$, where $\rho(x)$ means " $x$ is an element of $A$ " is such that for each finite family $(a_i)$ of elements in $A$, there is a $y$ in $A$ such that $\mathcal{R}(a_i, y)$ for each $i$. So $^*A$ <u>contains an element larger than every element</u> of $A$.

Thus, a part $A \subset \mathbb{R}$ is bounded if and only if each element of $^*A$ is finite (easy exercise).

<u>Order of procedure</u>. <u>Article 5</u>. Let $\mathcal{R}$ be a binary relation on $\mathbb{R}$ : Its source is the set of all $x$ in $\mathbb{R}$ for which there exists at least one $y$ such that $\mathcal{R}(x, y)$. We shall say that $\mathcal{R}$ <u>is of type</u> $\Gamma_2$ if and only if for each finite part $F$ of the source, there exists an element $y$ related to all $x$ in $F$.

<u>To invoque</u> $\Gamma_2$ is to hope that for any relation $\mathcal{R}$ of type $\Gamma_2$, there exists an $\omega$ in $^*\mathbb{R}$ related by the transfered relation to all elements of the source (in $\mathbb{R}$).

The proofs in § 5 and § 7 above suppose the hope already fullfilled for the given enlargement of $\mathbb{R}$.

From now on, we allow only the use of enlargements such that :

i) Their transfer * respects the well stated first order properties (i.e. with quantifications only on element -variables).

ii) The invocation of $\Gamma_2$ is justified.

Note that we could introduce a type $\Gamma_n$ for n-ary relations, $n > 2$ . But re-grouping the entries, we immediately reduce their number to two.

Comment. We may doubt whether any such enlargements exist at all. We could regard their existence as an axiom of mathematics. Of course, this could contradict some axiom of set theory.

Fortunately, the existence of enlargements follows from the axiom of choice (see section II).

We pose as a working hypothesis the existence of enlargements enjoying the properties stated up to now.

We shall add in due time further properties.

# Lesson 7

## MORE REAL TOPOLOGY

THEOREM. Let $A \subset \mathbb{R}$ . We have the following characterizations :

1) $A$ infinite if and only if $^*A \neq A$ ;

2) $A$ bounded $^*A \subset F$ ;

3) $A$ open $h(A) \subset {}^*A$ ;

4) $a$ interior to $A$ $h(a) \subset {}^*A$ ;

5) $a$ adherent to $A$ $h(a)$ meets $^*A$ ;

6) $a$ boundary point of $A$ $h(a)$ meets $^*A$ and $\complement_{^*\mathbb{R}} {}^*A$ ;

7) $a$ accumulation point of $A$ $h(a) - \{a\}$ meets $^*A$ ;

8) $f : A \longrightarrow \mathbb{R}$ continuous at $a$ $f(h(a)) \cap {}^*A \subset h(f(a))$ .

It is necessary to remember that : $A \subset {}^*A$ ;

- $F$ is the set of finite elements in $^*\mathbb{R}$ ;

- $h(A)$ is the halo of $A$ , i.e. the set of the elements of $^*\mathbb{R}$ that are infinitely close to some element in $A$ .

Applications. 1) Any infinite part $S$ contained in a closed bounded $A$ in $\mathbb{R}$ admits an accumulation point in $A$ .

Proof (that uses 1, 2, 5). Let $\alpha$ be in $^*S - S$ . As $^*A \subset F$ and $^*S \subset {}^*A$ , $\alpha$ is finite, thus it admits a standard part $a$ in $\mathbb{R}$ . As $h(a)$ meets $^*S$ , $a \in \overline{S} \subset \overline{A} = A$ . But $\alpha$ is not standard, thus it is distinct from $a$ .

2) $\mathbb{R}$ is connected.

Proof. Suppose that $\mathbb{R} = U \cup V$ where $U$ and $V$ are nonempty disconnected open sets: If $a$ is in $U$ and $b$ in $V$ , we have the double series with values in $[a , b]$

$$a_n^i = a + \frac{i(b-a)}{n} , \quad 0 \leq i \leq n .$$

For each $n$ there exists an integer $i(n)$ such that

$$a_n^{i(n)} \in U \quad \text{and} \quad a_n^{i(n)+1} \in V .$$

After transfer, we may fix $n$ as infinitely large, which gives two finite elements $\alpha$ and $\beta$, infinitely near, such that $\alpha \in {}^*U$ and $\beta \in {}^*V$. According to 3, $a$, the standard part common to these two points, would then be adherent to both $U$ and $V$, which are disconnected closed sets.

Comments. 1) Properties 2 to 8 are proved as an exercise using the existence of infinitesimals in ${}^*\mathbb{R}$ and very simple transfers.

2) On the other hand, property 1 is a consequence of the two following remarks :

i) if $A$ is finite, then the property $\forall x \in A$, $\exists i < n$ such that $x = a_i$ transfers itself into ${}^*A$. Now $\{i \in {}^*\mathbb{N}, i < n\} = \{i \in N, i < n\}$. Otherwise there exists $\omega \in {}^*\mathbb{N} - \mathbb{N}$ such that $\omega < n$. We shall see in lesson 8 that an element of ${}^*\mathbb{N} - \mathbb{N}$ is infinitely large, and so greater than $n$. As a result, ${}^*A = A$.

ii) If $A$ is infinite, the relation $\neq$ is of type $\Gamma_2$ because for each finite part $B \subset A$, $B - A$ is nonempty: As a result, ${}^*A - A$ is nonempty.

About the meaning of the word finite. We should distinguish between :

1) the intuitive or realistic notion of finite collection.

2) the axiomatic notion of finite set.

3) the notion of finite element in ${}^*\mathbb{R}$.

Notion 3 has certainly nothing in common with notions 1 and 2. Notions 1 and 2 may be considered as equivalent, provided we know how to distinguish one from the other. It is clear that the topological concept regards the finite n° 2. A better word for the "finite n° 3" would be "limited".

# Lesson 8

## FROM Q TO ℝ

THEOREM. From $^*\mathbb{N}$ we may extract Q .

Proof. Let n be infinitely large and prime in $^*\mathbb{N}$ ; the congruence classes, modulo n , constitute a field containing $\mathbb{N}$, thus also Q .

Comments. 1) Until now, we have considered $\mathbb{N}$ and Q as parts of $\mathbb{R}$ that have an enlargement in $^*\mathbb{R}$ . By restricting the transfer and the invocation of $\Gamma_2$ to elements, parts, functions, etc..., based on $\mathbb{N}$ , we obtain directly an enlargement of $\mathbb{N}$ . We can do the same for Q , or for any subset of $\mathbb{R}$ , "interesting" or not.

Thus there exist enlargements of $\mathbb{N}$ that are independent from the surrounding $\mathbb{R}$ .

2) Let us fix such an enlargement. According to the previous lesson, there exists in $^*\mathbb{N}$ an element $\omega$ different from all the elements of $\mathbb{N}$ . Such an element is also larger than all elements in $\mathbb{N}$ . Indeed, if $\omega < n$ , $n \in \mathbb{N}$ , the set of $x \leq \omega$ , x in $\mathbb{N}$ , is bounded by n , thus it possesses a greatest element $m_o$ ; we have then $m_o < \omega < m_o + 1$ ; thus $0 < \omega - m_o < 1$ . But, by transfer, there exists no element of $^*\mathbb{N}$ between 0 and 1 , since there exists none in $\mathbb{N}$ . Therefore there is in $^*\mathbb{N}$ an infinitely large element. This could be proved as well from the fact that the relation > is of type $\Gamma_2$ .

Notice that the finite elements of $^*\mathbb{N}$ , i.e. the non infinitely large ones are all standard (in $\mathbb{N}$ ).

If $\omega$ is infinitely large, so is $\omega \pm n$ for each n in $\mathbb{N}$ (see the joke on Darwin's monkeys !). Each $\omega$ generates a "galaxy" in bijection with Z .

The finite elements constitute a semi initial galaxy.

There is no smallest infinitely large (otherwise $\omega - 1$ is standard), no more than there is a greatest finite element (we can always add 1 ). Thus it is false that in $^*\mathbb{N}$ "each part has a smallest element" and that "each part that has an upper bound has a greatest element", and yet this looks like a transfer. This sort of mystery will be studied in greater detail in lesson 9. We must notice that if each part with an upper bound has a smallest element, we would deduce $^*\mathbb{N} = \mathbb{N}$ . The price to pay to have $^*\mathbb{N} \neq \mathbb{N}$ is the interdiction to transfer properties where we quantify <u>on parts</u>.

3) Here are some standard arithmetic theorems.

i) Addition and multiplication are associative, commutative, etc...

ii) The order relation is total and compatible with + and x .

iii) There is an euclidean division.

iv) If a divides bc and if a is prime with b , it divides c .

v) If n is prime, any element that is not a multiple of n possesses an inverse modulo n .

vi) For each a , there exists a prime greater than a .

vii) n! is divisible by a , whenever a inferior to n .

viii) Fermat's theorem : $a^{\varphi(n)} \equiv 1 \mod n$ , where $\varphi(n)$ is the number of number prime to n and less than n .

ix) If a property is true for 0 , and if $P(n)$ implies $P(n+1)$ , then P is true for all n .

x) The unique factorization theorem.

All these theorems are candidates to a transfer. From i) to viii) there is no problem because in their expression we quantify only on variables of elements in $\mathbb{N}$ . The proof of the introductive theorem relies on the transfer of v) and vi) .

Notice that : if $\omega$ is infinitely large, then $\omega!$ is divisible by all the elements of $\mathbb{N}$ . However we must restrain from taking the product in $^*\mathbb{N}$ of all elements of $\mathbb{N}$ ; for thus we would be dealing with the smallest number among all the infinitely large. (Further we will return to this point.)

4) Before transfering ix) , we should make P precise. If P is formulated by means of a sentence where just elements variables are quantified on $\mathbb{N}$ , everything is allright. Otherwise beware and proceed to lesson 9 !

5) Before transfering x) it has to be noticed that the factorization theorem is formulated as a forbidden quantification. It gives, half formalized : $\forall x$ , $\exists n$ and there exists a map $\lambda : [0, n] \longrightarrow \mathbb{N}$ such that $\forall i \leq n$ , $\lambda(i)$ is prime and $x = \lambda(0) \ldots \lambda(n)$ .

Two facts are opposed to the transfer :

- "there exists a map"

and

- the product of $\lambda(i)$ that is defined by recurrence on the finite parts of $\mathbb{N}$ (compare with $n!$ which is obtained by recurrence on the elements of $\mathbb{N}$ ). In order to transfer the unique factorization theorem, we would need the necessary authorizations. They will be given at lesson 9, and after we have been acquainted with the following order of procedure.

Order of procedure. About the properties of the first order.

Article 7. A property of the first order based on E is a formula of the set theoretical language where the only possible constants are, in addition to the elements of E , the parts of E but not of $\mathcal{P}(E)$ , the parts of $E \times E$ , etc..., and where for all the variables x that intervene, the formula contains " $x \in F$ " where F is one of those constants.

Example. $E = \mathbb{R}$ . The property $\forall x$ , $((x \in \mathbb{R}) \Longrightarrow \exists n \ (n \in \mathbb{N}) \text{ and } x > n))$ is of the first order on E .

Example. $E = \mathcal{P}(\mathbb{R}) \cup \mathbb{R}$ . The property

$\forall A$ , $(A \in \mathcal{P}(\mathbb{R})$ and $\exists x \ (x \in \mathbb{R}) \ldots)$

(or each enlarged part of $\mathbb{R}$ admits an upper bound) (in order to say that $y < x$ , we write that $(x, y)$ is element of the graph of the order relation on $\mathbb{R}$ ) is of the first order on $\mathbb{R} \cup \mathcal{P}(\mathbb{R})$ but not on $\mathbb{R}$ , or on $\mathcal{P}(\mathbb{R})$ .

Article 8. It is forbidden to transfer from $E$ to ${}^*E$ the properties that are not of the first order on $E$ .

Article 9. Point 8 may be dodged by the following trick : a property concerning $E$ and its parts is of the first order on $E \cup \mathcal{P}(E)$ . Therefore it transfers itself into ${}^*(E \cup \mathcal{P}(E))$ .

Article 10. Those who transgress point 8 by point 9 are asked to think on the following fact : if $F = \mathbb{N} \cup \mathcal{P}(\mathbb{N})$ , then by transfer of the property

$$\forall x \quad ((x \in F) \Longrightarrow (x \in \mathbb{N} \quad \text{or} \quad x \in \mathcal{P}(\mathbb{N}))$$

we have ${}^*F = {}^*\mathbb{N} \cup {}^*\mathcal{P}(\mathbb{N})$ . We saw above that the property "each part of $\mathbb{N}$ has a smallest element" does not apply to the parts of ${}^*\mathbb{N}$ , but by transfer into ${}^*F$ it applies to the elements of ${}^*\mathcal{P}(\mathbb{N})$ . So that $\mathcal{P}({}^*\mathbb{N}) \neq {}^*\mathcal{P}(\mathbb{N})$ .

---

## Lesson 9

## DIGRESSION ON UPPER-BOUNDED PARTS

COUNTER-THEOREM. "Some upper-bounded parts in ${}^*\mathbb{R}$ have no least upper bound".

THEOREM. Each internal and upper-bounded part of ${}^*\mathbb{R}$ has a least upper bound.

Comments. 1) The existence of infinitely large numbers shows that $\mathbb{R}$ has an upper bound in ${}^*\mathbb{R}$ . If $a$ is one of them, then so is $a - 1$ , which justifies the counter-theorem and, according to the theorem, $\mathbb{R}$ is not an internal part of ${}^*\mathbb{R}$ .

2) In the absence of a definition for "internal parts of ${}^*\mathbb{R}$ ", we

could call "internal" the parts with an upper bound that have a least upper bound. It comes to adjusting the definition to the theorems. Immoral...

3) The truth is suggested by point 10 (lesson 8) of the order of procedure. In an enlargement of $\mathbb{R} \cup \mathcal{P}(\mathbb{R})$, the elements of ${}^*\mathcal{P}(\mathbb{R})$ are related to parts of ${}^*\mathbb{R}$. We may indeed consider their belonging as a binary relation whose graph is a part of $\mathbb{R} \times \mathcal{P}(\mathbb{R})$. It transfers itself (with its properties) into a relation, temporarily written $\Subset$, between elements of ${}^*\mathbb{R}$ and elements of ${}^*\mathcal{P}(\mathbb{R})$.

4) Let $A$ belong to ${}^*\mathcal{P}(\mathbb{R})$. To $A$ we may associate the set $\hat{A}$, $\hat{A} = \{x \in {}^*\mathbb{R} , x \Subset A\}$ which is itself an element of $\mathcal{P}({}^*\mathbb{R})$.

The mapping $A \longrightarrow \hat{A}$ so defined is an injection of ${}^*\mathcal{P}(\mathbb{R})$ in $\mathcal{P}({}^*\mathbb{R})$. Thus, if $\hat{A} = \hat{B}$, we have the property

$$\forall x \in {}^*\mathbb{R} , \quad x \Subset A \Longleftrightarrow x \Subset B .$$

By transfer of the property of extensionality into $\Subset$, we deduce that $A = B$.

The internal parts of ${}^*\mathbb{R}$ are now the parts of the form $\hat{A}$, with $A \in {}^*\mathcal{P}(\mathbb{R})$. We may abusively consider ${}^*\mathcal{P}(\mathbb{R})$ as a part of $\mathcal{P}({}^*\mathbb{R})$.

Notice that, if $A \in \mathcal{P}(\mathbb{R})$, ${}^*A$ defines an internal part. But there exist internal parts that are not of this type (see 9).

5) The property (of the second order) on $\mathbb{R}$ : "each part that has an upper bound has a least upper bound" is of the first order on $\mathbb{R} \cup \mathcal{P}(\mathbb{R})$, therefore it transfers itself into ${}^*(\mathbb{R} \cup \mathcal{P}(\mathbb{R}))$. It becomes "each internal part that has an upper bound has a least upper bound".

Thus $\mathbb{R}$, ${}^*\mathbb{R} - \mathbb{R}$, the set of finite elements, infinitesimal, etc..., are "external" parts because they do not verify this property.

6) The internal parts verify all the properties of the first order based on $\mathbb{R} \cup \mathcal{P}(\mathbb{R})$ and transfered into ${}^*\mathbb{R} \cup {}^*\mathcal{P}(\mathbb{R})$.

7) Concerning elements of ${}^*\mathbb{R}$, there is no difference between elements and internal elements. This explains the restriction to the first order of

the properties that we can transfer with no further precaution.

8) If we want to transfer properties where we quantify on sets of functions, relations, ... , it is necessary to consider an enlargement of the corresponding type of object. If we want <u>all</u> freedom, we must enlarge

$$E = \mathbb{R} \cup \mathcal{P}(\mathbb{R}) \cup \mathcal{P}(\mathbb{R}\times\mathbb{R}) \cup \ldots \cup \mathcal{P}(\mathbb{R}^i) \cup \ldots \mathcal{P}(\mathcal{P}(\mathbb{R}))\cup\mathcal{P}(\mathbb{R}\times\mathcal{P}(\mathbb{R})) \cup \ldots$$

We swiftly become dizzy ; in practice, we use only a slight part of this <u>complete structure</u> built on $\mathbb{R}$ , whose precise definition is made with the help of the "scale of types" : the set $T$ of all types is the smallest set such that $0 \in T$ and that, for each finite sequence $\tau_1, \ldots, \tau_n$ of elements of $T$ , the n-uple $(\tau_1, \ldots, \tau_n) \in T$ .

For each $\tau \in T$ , we define a set $\mathbb{R}_\tau$ by the recurrence relations

$$\mathbb{R}_0 = \mathbb{R} ,$$

$$\mathbb{R}_{(\tau_1,\ldots,\tau_n)} = \mathcal{P}(\mathbb{R}_{\tau_1} \times \ldots \times \mathbb{R}_{\tau_n}) .$$

Then $E = \bigcup_{\tau\in T} \mathbb{R}_\tau$ .

In an enlargement of $E$ , we shall have one ${}^*\mathbb{R}_\tau$ for each $\tau$ ; if $\tau = (\tau_1, \ldots, \tau_n)$ , we may compare ${}^*\mathbb{R}_\tau$ with $({}^*\mathbb{R}_{\tau_1} \times \ldots \times {}^*\mathbb{R}_{\tau_n})$ when considering the relation "$\lessdot$" transfered from "$\in$" between elements and parts of $\mathbb{R}_{\tau_1} \times \ldots \times \mathbb{R}_{\tau_n}$ . As in (4), the result is a natural injection that defines the internal parts of ${}^*\mathbb{R}_{\tau_1} \times \ldots \times {}^*\mathbb{R}_{\tau_n}$ . For example, an internal function of ${}^*\mathbb{R}$ in ${}^*\mathbb{R}$ is obtained from an element of ${}^*\mathcal{P}(\mathbb{R} \times \mathbb{R})$ .

9) In the complete structure $E$ , we can transfer unions, intersections, complement, projections. Thus these operations transform internal objects into external objects.

Consider a predicate with one free variable $A(x)$ of which all the ingredients, except for the group symbols, the quantifiers, the logical connectives and the variables, are elements of ${}^*E$ . Then $\{x \in {}^*\mathbb{R} , A(x)\}$ is internal, because it is obtained by combination of the above-mentioned set operations.

10) Here is an exercise based on this remark.

*Let* $I$ *be a part of* $\mathbb{R}$ . *Then* $I$ *is internal as a part of* ${}^*\mathbb{R}$ *if and only if it is finite*.

We suppose that we enlarge $\mathbb{R} \cup \mathcal{P}(\mathbb{R}) \cup \mathcal{P}(\mathbb{R} \times \mathbb{R})$ : If $I$ is finite, we have ${}^*I = I$ and $I$ is internal. If $I$ is infinite, there exists an injection $f : \mathbb{N} \longrightarrow I$ and by transfer, an injection ${}^*f : {}^*\mathbb{N} \longrightarrow {}^*I$ . As ${}^*f$ and ${}^*\mathbb{N}$ are internal, if $I$ is internal, the set : $\{x \in {}^*\mathbb{N} \ , \ {}^*f(x) \in I\}$ is also internal. It contains $\mathbb{N}$ that is not internal, as we know. Thus there exists $\omega \in {}^*\mathbb{N} - \mathbb{N}$ such that ${}^*f(\omega) \in I$ . Then

$$ {}^*f(\omega) \in {}^*f({}^*\mathbb{N}) \cap I = {}^*(f(\mathbb{N})) \cap I \subset {}^*(f(\mathbb{N})) \cap \mathbb{R} = f(\mathbb{N}) \ , $$

so that $n \in \mathbb{N}$ with ${}^*f(n) = {}^*f(\omega)$ does not exist, which interferes with the injectivity of $f$ .

11) To point at the existence of an element $\omega$ with a certain property by stating that "the set of the elements with the property is internal and contains $\mathbb{N}$, and thus is not reductible to $\mathbb{N}$ " is a rewarding procedure. We shall see it often at work later on.

# Lesson 10

## INTERNAL SEQUENCES

Definition 1. An internal sequence $u$ on ${}^*\mathbb{R}$ is $*$-convergent if and only if there exists $\ell \in {}^*\mathbb{R}$ such that for each $\varepsilon > 0$ in ${}^*\mathbb{R}$, there exists $n_o \in {}^*\mathbb{N}$ such that $n > n_o$ implies $|u_n - \ell| < \varepsilon$.

Definition 2. An internal sequence $u$ on ${}^*\mathbb{N}$ is S-convergent if and only if there exists $\ell \in {}^*\mathbb{R}$ finite, such that, for each $n$ infinitely great in ${}^*\mathbb{N}$, $u_n \sim \ell$.

Proposition 1. If $u$ is S-convergent, there exists $\alpha \in \mathbb{N}$ such that for $n \geq \alpha$, $u_n$ is finite and the standard sequence ${}^o(u_n)$ (with arbitrary values for $n < \alpha$) converges in $\mathbb{R}$ towards ${}^o\ell$.

Proposition 2. A sequence of type ${}^*u$ is S-convergent, if and only if there exists a finite $\ell \in {}^*\mathbb{R}$ and an infinitely large $\omega$ such that ${}^*u_n \sim \ell$ for every infinitely large $n$ up to $\omega$.

Definition 3. An $*$-finite sequence $u_1, \ldots, u_\omega$ $(\omega \in {}^*\mathbb{N})$ is S-convergent if there exists a finite, $\ell \in {}^*\mathbb{R}$, such that $u_n \sim \ell$ for every infinitely large $n$, up to $\omega$.

Comments. 0) The above definitions make sense in an enlargement of $\mathbb{R} \cup \mathcal{P}(\mathbb{R} \times \mathbb{R})$. An internal sequence is a map of ${}^*\mathbb{N}$ into ${}^*\mathbb{R}$ whose graph is an element of ${}^*\mathcal{P}(\mathbb{R} \times \mathbb{R})$. For example, if $u : \mathbb{N} \longrightarrow \mathbb{R}$ is a standard sequence, ${}^*u$ is an internal sequence. But the sequence $u_n = \frac{\omega}{n}$, with $\omega$ infinitely large, is internal although not of type ${}^*u$ because for $n \in \mathbb{N}$, $u_n \notin \mathbb{R}$.

1) By transfer, a sequence of type ${}^*u$ is $*$-convergent iff $u$ is convergent. According to lesson 1, a sequence of type ${}^*u$ is S-convergent iff $u$ is convergent. In the case of those sequences, both notions coincide. But the se-

quence $\frac{\omega}{n}$ is $*$-convergent although not S-convergent because $\frac{\omega}{\omega} = 1$ and $\frac{\omega}{2\omega} = \frac{1}{2}$ .

On the other hand, an internal sequence may be S-convergent and not $*$-convergent, for instance $(-1)^n \varepsilon$ with $\varepsilon \sim 0$ .

2) According to proposition 1, an internal S-convergent sequence is "almost" of type ${}^*V$ where $v_n = {}^\circ(u_n)$ for $n \geq \alpha$ , because ${}^*v_n \sim {}^\circ\ell \sim u_n$ for n infinitely large. The difference $v_n - u_n$ is an internal sequence with infinitesimal values.

3) We prove proposition 1 by means of a very simple idea called "principle of permanence", or "Robinson's lemma".

If U is S-convergent, there exists a number K in $\mathbb{R}$ such that $\ell < K$ . Then, the set $\{p \in {}^*\mathbb{N} , n > p \Longrightarrow |U_n| < K\}$ is internal. It contains all the infinitely large, thus it contains also $\alpha \in \mathbb{N}$ (the set of the i.g. is not internal because it has no smallest element). The sequence ${}^\circ u_n$ from $\alpha$ on. Let $\varepsilon$ be $> 0$ in $\mathbb{R}$ ; for n infinitely large, we have $|u_n - \ell| < \varepsilon$ : this property is permanent, up to a standard $n_o$ , so that ${}^\circ u_n$ tends to ${}^\circ\ell$ , because $|{}^\circ u_n - {}^\circ\ell| \sim |u_n - \ell|$ .

The converse is immediate, if we consider the limit characterization.

4) Proposition 2 illustrates a new fact ; the convergence of u depends only on the values of ${}^*u$ up to an infinitely large $\omega$ . This takes on its full flavour if we consider " $*$-finite" sequences, cut off at $\omega$ ; the notion of convergence in definition 3 provides a quite adapted frame for the formulation of discrete probability laws in terms of finite spaces of probability (Read the presentation of the laws of the great numbers in E. Nelson [N], for example).

Suppose then that ${}^*u_n \sim \ell$ for each $n \leq \omega$ , n infinitely large. Let $\varepsilon > 0$ in $\mathbb{R}$ . Under the principle of permanence, there exists $n_o \in \mathbb{N}$ such that for $n_o < n \leq \omega$ , $|{}^*u_n - {}^\circ\ell| < \varepsilon$ , so that for $n \in \mathbb{N}$ , $n > n_o$ , we have $|u_n - {}^\circ\ell| < \varepsilon$ and the sequence ${}^*u$ is S-convergent.

5) Use the principle of permanence to show that an internal sequence u such that $u_n \sim 0$ for each $n \in \mathbb{N}$ verifies $u_n \sim 0$ up to an i.l. $\omega$ . (Look at the property $|u_n| < \frac{1}{n}$ .)

6) We may discuss in a similar way the limit of internal functions ; *-limit, S-limit, *-continuity and S-continuity, with the use of a principle of permanence analogous to that previously described.

---

## Lesson 11

## FROM $^*\mathbb{Q}$ WE MAY EXTRACT $\mathbb{R}$

THEOREM. Let K be a commutative field, totally ordered and archimedean. In an enlargement of K .

1) The set F of the finite elements of $^*K$ is a K-subalgebra of $^*K$ containing K .

2) The set I of the infinitesimal elements of $^*K$ is an ideal of $^*K$ .

3) The K-algebra $\tilde{K} = F/I$ is a commutative totally ordered complete archimedean field containing K up to an isomorphism.

COROLLARY. K is complete if and only if each finite element of $^*K$ is in the halo of an element of K .

Comments. 1) By transfer, $^*K$ is a commutative and totally ordered field, $^*\mathbb{N}$-archimedean. As K is $\mathbb{N}$-archimedean, the order relation is of type $\Gamma_2$ , which ensures the existence of a $\omega$ in $^*K$ larger than every element of K . Thus $^*K$ contains :

- the set K of standard elements.
- the set F of finite elements (less in absolute value than some

standard element).

- the set $K \setminus F$ of infinitely large elements (larger in absolute value than every element of $K$).

- the set $I$ of infinitesimal elements (less in absolute value than every positive standard element).

2) The computation rules in the $K$-algebra ${}^*K$ are the following :

(Infinitely large) x (Infinitely large) = (Infinitely large) ;

(Finite not infinitesimal) x (Infinitely large) = (Infinitely large) ;

(Finite) + (Infinitely large) = (Infinitely large) ;

(Positive infinitely large) + (Positive Infinitely large)

= (Positive infinitely large) ;

(Negative infinitely large) + (Negative infinitely large) =

= (Negative infinitely large) .

From this we deduce the computation rules on the infinitesimals, such elements being either zero or the inverse of an infinitely large one. From there we prove points 1 and 2 of the theorem.

3) <u>Proof of theorem 3</u>. The total order of ${}^*K$ restricted to $F$ is compatible with the quotient by $I$ and the one to one map $K \longrightarrow F$ defines a one to one map $K \longrightarrow \tilde{K} = F/I$ . $K$ and $\tilde{K}$ being ordered ring, it defines a total order on $\tilde{K}$ .

- $\tilde{K}$ is archimedean. Indeed if $\alpha \in F$ , there exists $k$ in $K$ such that $|\alpha| < k$ . As $K$ is archimedean, we may suppose that $k$ is an integer. Thus $cl(\alpha) \leq cl(k)$ .

- $\tilde{K}$ is a field. Let $\alpha$ be in $F$ and $\alpha \notin I$ . Then $\frac{1}{\alpha} \in F$ and $cl(\alpha) \cdot cl(\frac{1}{\alpha}) = 1$ :

- $\tilde{K}$ is complete. Indeed $Q$ is dense in $K$ , because it is archimedean. Then $K$ is dense in $\tilde{K}$ . Let then $(u_n)$ be a sequence of Cauchy of $\tilde{K}$ . There exists a sequence $(v_n) = cl(\alpha_n)$ with $\alpha_n$ in $K$ such that $|u_n - v_n| \leq \frac{1}{n}$ . Let $\varepsilon > 0$ , $\varepsilon \in \tilde{K}$ ; $\exists\, \eta > 0$ , $\eta \in K$ such that $cl(\eta) < \varepsilon$ . The sequence $v_n$ being a

Cauchy sequence, there exists $n_o \in \mathbb{N}$ such that for $p, q > n_o$, we have $|cl(\alpha_p) - cl(\alpha_q)| < cl(\eta)$ ; i.e. $|\alpha_p - \alpha_q| < \eta$ + inifnitesimal, which is equivalent to $|\alpha_p - \alpha_q| \leq \eta$ (the difference between two distinct elements of K cannot be infinitesimal).

Let $\omega$ be an infinitely large integer. We have $|\alpha_p - {}^*\alpha_\omega| \leq \eta$ (transfer). Therefore ${}^*\alpha_\omega$ is finite. Let $\ell = cl({}^*\alpha_\omega)$ . We have $|v_p - \ell| < \varepsilon$ for $p > n_o$ which shows that $\ell$ is the limit of $u_p$ .

4) We can prove directly that $\tilde{K}$ is complete in the sense of the upper bounds by using an enlargement of $K \cup \mathcal{P}(K)$ .

5) The unicity of commutative fields totally ordered, complete and archimedean shows that K is isomorphic to $\mathbb{R}$ , in particular $\tilde{Q} \simeq \mathbb{R}$ , which justifies the title of the lesson.

6) The equivalence relation modulo I extends itself to the whole of ${}^*K$ , but it is not compatible with multiplication.

7) Proof of the corollary. See lesson 0 .

PLAY-TIME

1) Let $c(t)$ be a curve in $\mathbb{R}^2$. Prove that it is the shadow of some curve in ${}^*\mathbb{R}^2$ without any cusp. Also prove that a singular point of first (resp. second) kind is the shadow of a singular of second (resp. first) kind.

2) Find an $*$-analytic curve in ${}^*\mathbb{R}^2$ whose shadow is the unit square in $\mathbb{R}^2$.

3) Find an $*$-analytic surface in ${}^*\mathbb{R}^3$ whose shadow is the unit cube in $\mathbb{R}^3$.

4) Prove that any parabola in $\mathbb{R}^2$ is the shadow of some $*$-ellipse in ${}^*\mathbb{R}^2$.

Prove that it is also the shadow of some $*$-hyperbola.

May an hyperbola be the shadow of some $*$-ellipse (use the polar equation $\rho = \dfrac{1}{1 + e\cos\theta}$ to define $*$-conics with non standard excentricity $e$) ?

5) Prove that a circle is the shadow of an $*$-ellipse.

6) Let $E$ be an $*$-ellipse whose shadow is a circle $C$.
An $*$-isometry of ${}^*\mathbb{R}^2$ is called an "almost symmetry" of $E$ if the image of each point in $E$ is infinitely near some point of $E$.
Relate these almost symmetries with the symmetries of $C$ and keep in mind that physicists often have to do with quick discontinuities in the symmetry groups of evolutive phenomena.

7) Let $f : {}^*\mathbb{R} \longrightarrow {}^*\mathbb{R}$ be the function $\sin \pi\omega!\, x$ with $\omega$ an infinitely large integer. Prove that the shadow of $f$ exists and vanishes on all rational numbers. May it take the values $+1$ or $-1$ ?

8) Find an internal function on ${}^*\mathbb{R}$ whose shadow is continuous and has a derivative nowhere. May this function have a derivative ?

9) Let A be a complex $n \times n$ matrix. Prove that A is the shadow of some $*$-matrix with all its eigenvalues distinct.

10) Let A be a real $n \times n$ matrix, with all its eigenvalues distinct and real. Prove that any $*$-matrix B with $°B = A$ has all its eigenvalues real and distinct.

11) Prove that in some precise sense, Euclidean geometry is the shadow of hyperbolic geometry in two dimensions. Compare the metrics using the model of Poincaré.

12) Let $\mathfrak{g}$ be some real Lie algebra of dimension 3. Consider an $*$-linear mapping $f : {}^*\mathbb{R}^3 \longrightarrow {}^*\mathbb{R}^3$ and for all $x, y$ in $\mathbb{R}^3$ with $f^{-1}[f(x), f(y)]$ finite, put $[[x, y]] = °(f^{-1}[f(x), f(y)])$.
Prove that, whenever $[[\ ]]$ is defined everywhere on $\mathbb{R}^3$, it is a Lie algebra bracket. We write $\mathfrak{g}_o$ for the corresponding algebra and call it a deformation of $\mathfrak{g}$.

13) Let $\mathfrak{g}_o$ be a deformation of $\mathfrak{g}$ as above. Assume that there is in $\mathfrak{g}_o$ a linear form $\omega$ such that $\omega \wedge d\omega \neq 0$ (invariant contact form on the corresponding Lie group). Prove that this property is also true for $\mathfrak{g}$. (Recall that the differential d is associated with the Lie bracket through the formula $d\omega(x, y) = -\omega([x, y])$.

14) Consider the $*$-differential equation $\varepsilon\dot{x}(t) + x(t) = t$ in ${}^*\mathbb{R}$, with $\varepsilon \sim 0$. An equivalent system is
$$\begin{cases} \dot{x} = \frac{1}{\varepsilon}(t - x) \\ \dot{t} = 1 \end{cases}$$

- Prove that the $*$-integral curves starting at a finite point outside the halo H of the curve $x = t$ are almost horizontal as long as they don't meet this halo.
- Prove that after some infinitesimal time, such a curve reaches some point in this halo.
- Prove that the $*$-integral curves starting at a finite point of H remain in H for every finite time.

• Prove that if $t_o > x_o$, the integral curve starting at the finite $(t_o, x_o)$ crosses the curve $x = t$ near $(t_o, t_o)$.

• Prove that if $t_o < x_o$, the integral curve starting at the finite point $(t_o, x_o)$ does not cross the curve $x = t$.

Challenge : Do all this without any computation.

• Compute the integral curves and compare with the qualitative study.

• Imagine an other equation where computation is hopeless, but where your qualitative observations still work.

15) Consider a polynomial $P_o = ax^2 + bx + c$ in $\mathbb{C}[x]$ and an $*$-polynomial $P = \alpha x^2 + \beta x + \gamma$, with $\alpha, \beta, \gamma \in {}^*\mathbb{C}$ such that $\alpha \sim a$, $\beta \sim b$, $\gamma \sim c$. Assume $a \neq 0$. Prove that the roots of $P_o$ are the shadows of the roots of $P$.

16) Consider the $*$-polynomial $P = \varepsilon x^3 + x^2 - 3x + 2$ with $\varepsilon \in {}^*\mathbb{R}$, $\varepsilon \sim 0$. Prove that two roots of $P$ have as shadows 1 and 2 and that the third root is infinitely large.

HINTS

1) Use ${}^*c(t) + (\varepsilon t, 0)$, with $\varepsilon \sim 0$. This effaces an isolated singularity.

By first kind, we mean that the curve $(\dot{x}(t), \dot{y}(t))$ is regular near the singular point (without inflexion). The second kind has inflexion. Add something to change the kind.

2) Use $x^{2\omega} + y^{2\omega} = 1$ with $\omega$ infinitely large.

3) See (2) and generalize.

4) Add some infinitesimal to $e = 1$, with different signs and compute the shadows.

5) Don't forget to prove that every point on the circle is a shadow.

6) Every $*$-isometry which keeps the center of $E$ (and $C$) fixed is

an almost symmetry of E . The symmetries of C are precisely the shadows of these $*$-isometries. Notice that the $*$-symmetries of E (i.e. $*$-isometries which preserve globally E ) are only the four elements of a Klein group. This is a very simple example of singular perturbation : a family of finite groups ending with a continuous group as the parameter reaches some special value.

7) If $x \in Q$ , $\omega!\,x$ is an integer for $\omega!$ is a multiple of any standard integer.

8) Take $g(x) = \sum_{n=0}^{\omega} (\frac{2}{3})^n \cos 15\pi x$ with $\omega$ infinitely large.

9) Use a triangular form of A and add some infinitesimals on the diagonal.

10) If you don't succeed, see Lesson IV.2.

11) The model of Poincaré is the open upper half plane in $\mathbb{R}^2$ with the following "hyperbolic" straight lines :

- all half-circles whose center is on the x-axis.
- all vertical euclidian half-straights.

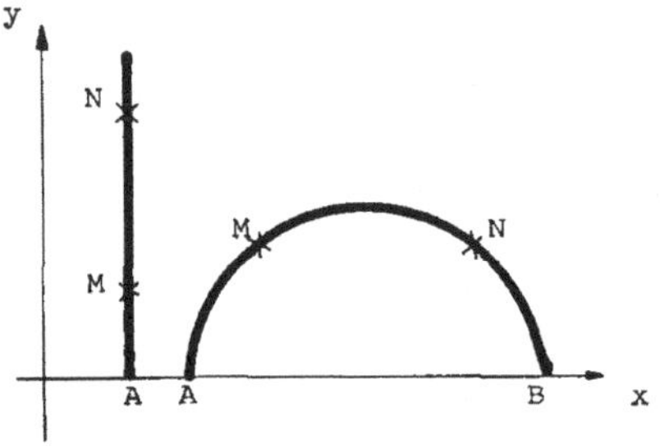


The distance between two points M and N is :

$d(M , N) = \log\left|\frac{m - a}{m - b} : \frac{n - a}{n - b}\right|$ ,

where $(a , b , m , n)$ is the complex representation of $(A , B , M , N)$ (in the degenerate case, consider that $\frac{n - b}{m - b} = 1$ .)

Now, enlarge this structure and consider the mapping $f : {}^*\mathbb{R}^2 \longrightarrow {}^*\mathbb{R}^2$ defined by $f(X , Y) = (X , Y+\omega)$ , where $\omega$ is infinitely large $> 0$ . Pull back the hyperbolic $*$-lines by f and consider their shadows in the first ${}^*\mathbb{R}^2$ .

12) and 13) If you don't succeed, see lesson IV.4.

14) The vector field associated with the system is nearly horizontal in the finite plane, except in H . Use general properties of vector fields to con-

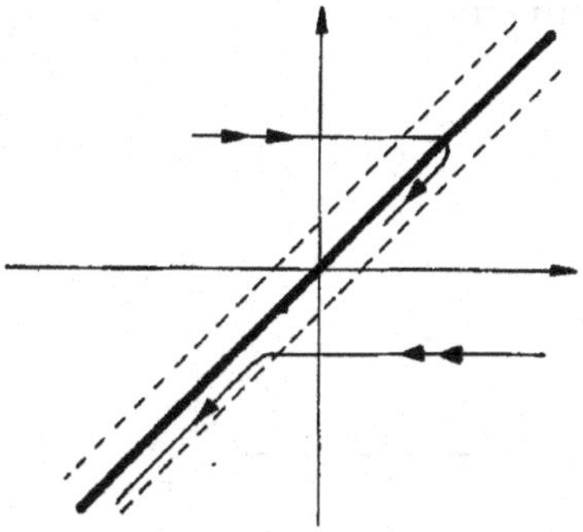

clude outside H .

To reach a point in H , use the permanence principle. Other points are due to the behaviour of the field along the line $x = t$ .

15) Compute the roots. Try also to answer without any computation.

16) Use the relations between roots and coefficients.

---

PART II : LOGICAL FOUNDATIONS OF NON - STANDARD ANALYSIS

## Lesson 1

## A REVIEW OF THE FOUNDATIONS : Z.F.

THEME. The formal theory Z.F. is a basis for classical mathematics.

Comments. 0) Z.F. is the canonical abbreviation for Zermelo-Frankel, the authors of the well-known axiom system that is commonly considered as an adequate formalization of the intuitive set concept.

Of course, an up-to-date mathematician will learn nothing in this lesson, for he works in a frame which could be formalized within Z.F.

In fact, much math-makers are not especially interested in foundations and the "sets" they have in mind are rather informal collections. But here we have to justify N.S.A. and this needs a little more formalism.

Clearly, Z.F. is a long story and we cannot tell it in full detail here. Our aim is rather to harmonize our points of view and to clarify the essential facts we need in our discussion. Consider this lesson as a pleasure-walk through set theory...

1) The usual mathematical language partly is a derivative of natural languages which vocabulary is given a new meaning (groups, rings, fields, ...), partly uses specific words (diffeomorphism, ...) created for some purpose. Long informal discussions and well admitted traditions are needed in order to clarify the use of this language.

The last refinement of these discussions leads to a formal language : all terms loose their intuitive sense and one retains only the syntactic form of the sentences about these terms. The ingredients of a formal language are signs, that is material figures, which may be immortalized with a graver on a marble plate - or, for want of marble, on a camel shoulder bone, but these modern technics give less durable products, alas ! Any clever copyst should be able to recognize such signs and to reproduce them without alteration, as often as necessary.

2) The specific signs of $\mathcal{L}_{Z.F.}$, the language of the formal theory Z.F. are $\in$ , $=$ , $\vee$ , $\neg$ , $\exists$ . As in any mathematical language, there are also unspecific signs, the <u>variables</u>, usually taken in familiar alphabets (latin, greek, gothic...) - the chinese one would be of help, but we don't master it...

The use of variables is supposed to be unlimited.

3) Writing these signs in some order along the lines of a copy plate, we may produce complicated <u>assemblages</u> which only are bounded by our imagination and the cost of marble...

But most of them are not useful and even not acceptable if we want to interprete informally (i.e. with our subjective mind) the sentences of $\mathcal{L}_{Z.F.}$ in the intuitive set theoretic language. For instance $x \in = y$ has to be rejected, for our intuitive language has grammatical rules which assign different plays to its ingredients : here $\in$ and $=$ should be translated as verbs and we never have two verbs for the same subject and complement.

The simplest assemblages we need are $= x\ y$ and $\in x\ y$ where $x, y$ are arbitrary variables (we write first without using brackets and hence put the "verb" before the subject) ; call them <u>atomic formulas</u>.

From atomic formulas, we obtain other <u>formulas</u> by repeated use of the following operations :

(n) put $\neg$ before a formula

(o) put $\vee$ before a sequence of two formulas

(q) put $\exists$ before a variable followed by a formula.

This is the grammar of $\mathcal{L}_{Z.F.}$ : the formulas are the well-formed assemblages following these rules. For instance $\exists\, x \vee \vee \in x\ y = xy \neg = xy$ is a formula because we can describe its construction in four steps :

$$\left.\begin{array}{l} = xy \longrightarrow \neg = xy \\ \left.\begin{array}{l}\in xy \\ = xy\end{array}\right\} \longrightarrow \vee \in xy = xy \end{array}\right\} \longrightarrow \vee \vee \in xy = xy \neg = xy$$

$$\longrightarrow \exists\, x \vee \vee \in xy = xy \neg = xy :$$

But $x \exists \vee \in \vee xy$ is not a formula.

Note that we cannot give precise rules of incorrectness, which is not important, for we are not in want of incorrect formulas...

The collection (please don't call it a "set"...) of all formulas is <u>potentially infinite</u>, because repeating operation (n) , (o) , (q) doesn't alter them ! The simplest well known potentially infinite collection is obtained by the rule "put | behind" : you get the intuitive integers | , || , ||| , |||| , ||||| , etc... (read phonetically one, two, three, four, five...) : whether this collection has something to do with the "set $\mathbb{N}$ of integers" or not is an important point in our further discussion.

4) In order to define or prove some property of the formulas, one has to follow their fabrication programm. For instance, define the <u>free variables</u> of a formula by the following rules :

- both variables in an atomic formula are free ;
- the free variables in $\neg A$ are the same as in $A$ ;
- the free variables in $\vee A B$ are those of $A$ and of $B$ ;
- the free variables in $\exists x A$ are the same as in $A$ , except $x$ if $x$ occurs in $A$ .

A formula without free variables is called <u>closed</u>. All its variables occur at least once in <u>quantified</u> form $\exists x$ .

We say that this property is defined by <u>finitary induction</u> ; any statement about freeness of variables has to be proved by finitary induction, in which each step and the transition between consecutive steps have to be material evidences.

David Hilbert and his scholars tried to clarify finitary principles that could be accepted by all mathematician. These principles are essential in both attitudes about the foundations of mathematics :

- in the combinatorial approach, the objects - e.g. the integers - are defined by finitary induction and their properties proved in the same way.
- in the formalistic approach, the language and the internal logic are

both described by finitary induction.

5) Of course, our language $\mathcal{L}_{Z.F.}$ could'nt be practised and related with intuitive concepts without abbreviations. In some sense, the usual mathematical language may be considered as an informal abbreviate of $\mathcal{L}_{Z.F.}$ ...

A formal abbreviation convention is often given by a scheme where the letters may be replaced by variables or formulas according to each case. Here are some examples :

replace $\in xy$ by $(x \in y)$

" $= xy$ by $(x = y)$

" $\neg = xy$ by $(x \neq y)$

" $\neg \in xy$ by $(x \notin y)$

" $\vee AB$ by $(A \vee B)$

" $\vee \neg A B$ by $(A \Longrightarrow B)$

" $\neg \vee \neg A \neg B$ by $(A \wedge B)$

" $(A \Longrightarrow B) \wedge (B \Longrightarrow A)$ by $A \Longleftrightarrow B$

" $\neg \exists x \neg A$ by $\forall x A$

The use of brackets makes $\mathcal{L}$ easier to translate but a computer would not need them. Of course, we are not computers...

We write $A(x_1, \dots, x_n)$ if it is useful to specify the free variables $x_1, \dots, x_n$ occuring in a formula $A$ .

6) Logicians have discovered that our usual logical deductions in mathematics can be reduced to the repeated use of the following rules :

- from $A$ deduce $(B \vee A)$
- from $A \vee A$ deduce $A$
- from $A \vee (B \vee C)$ deduce $(A \vee B) \vee C$
- from $(A \vee B)$ and $((\neg A) \vee C)$ deduce $B \vee C$
- if $x$ is not a free variable in $B$ , deduce $(\exists x A \Longrightarrow B)$ from $(A \Longrightarrow B)$ .

We recall these rules only to make the concept of logical deduction clear ; interested readers should make themselves familiar with logics in some specialized treatise.

Now, from an initial collection of formulas called axioms of the theory Z.F. (or axiom schemes where free entries may be replaced by any formula), we get by logical deduction the potential collection of all theorems of Z.F. Again we have no criteria to characterize those formulas which are not theorems, but this is not important.

For instance, we know that the addition of intuitive integers - we call them natural integers in the sequel - is commutative ; in the theory Z.F., there is a corresponding statement about the set $\mathbb{N}$ , which can be deduced from the axiom system. But in the first case, the property has to be proved by finitary induction, which is a completly different procedure. Indeed, both properties concern entities of different nature, although an exterior observer may find some informal relation between them.

The axioms and axiom schemes of Z.F. are :

- "logical" axioms concerning "equality" and "membership" :

  - $x = x$
  - $(((x_1 = y_1) \wedge (x_2 = y_2) \wedge (x_1 = x_2)) \Longrightarrow (y_1 = y_2))$
  - $((x_1 = y_1) \wedge (x_2 = y_2) \wedge (x_1 \in x_2)) \Longrightarrow (y_1 \in y_2)$ ;

- the schemes $((\neg A) \vee A)$ , where $A$ may be any formula without free variable and $((\forall x B) \Rightarrow B)$ where $x$ is a free variable occuring in a formula $B$ .

- the specific axioms which are quite popular ; we don't recall them. Notice that we don't consider the axiom of choice in Z.F., for the moment.

7) It is clear that these axioms should reflect as faithfully as possible the intuitive concept of set, which more or less is an idealisation of what we observe on material collections. However, at this time, $\mathcal{L}_{Z.F.}$ reflects nothing more than its grammar ! We have to translate it in our informal language. The basic lexicon is the following :

translate x in "the set x "

" $x \in y$ in "the set x is an element of the set y "

" $x = y$ in "the set x equals the set y "

" $\vee$ in "or"

" $\wedge$ in "and"

" $\neg$ in "not"

" $\Longrightarrow$ in "implies"

" $\Longleftrightarrow$ in "is equivalent"

" $\exists\, x$ in "there exists a set x such that..."

" $\exists\,!\, x$ in "there exists an unique set x such that..."

" $\forall\, x$ in "for every set x ...".

Now our theory Z.F. seems to tell us something about entities called "sets". Yet there is nothing in Z.F. but the syntactical material ! However, there are theorems like $\exists\,!\,x\, A(x)$ where x is a free variable in A . We would like to say that property A defines a set which should receive a specific name ; but we have no "names" in $\mathcal{L}_{Z.F.}$ ... Therefore, we introduce some constant a , with the grammatical rule " a may replace any free variable in any formula" and get a new language $\mathcal{L}'_{Z.F.}$ ; now introduce the supplementary axiom $A(a)$ and the axiom scheme $\forall\, y \;\; B(y, \ldots) \Longrightarrow B(a, \ldots)$ .

It is easy to check that in such an <u>extension by definition</u> Z.F' of Z.F. , for every formula T' in $\mathcal{L}'$ , there is a formula T in $\mathcal{L}$ such that $T \Longleftrightarrow T'$ is a theorem in Z.F'. ; moreover the extension is <u>conservative</u>, that is a formula in $\mathcal{L}$ is a theorem in Z.F. if and only if it is a theorem in Z.F'.

Thus there is no reason to avoid such extensions. Usually we call this procedure "set construction".

For instance $\exists\,!\,x\,\forall y\,(y \notin x)$ is a theorem in Z.F. Hence there is an extension where the constant $\emptyset$ is defined by $\forall y\,(y \notin \emptyset)$ . In a second extension, we get the constant $\{\emptyset\}$ defined by $\forall y\,(y \in \{\emptyset\}) \Longleftrightarrow (y = \emptyset)$ , because $\exists\,!\,z((\forall y\,(y \in z) \Longleftrightarrow (y = x))$ is a theorem in Z.F.

8) Let us admit that the reader, following the mathematical practice, defines himself $\cup$ , $\cap$ , $\subset$ , $\mathcal{P}(x)$ , etc... and consider the fundamental sets $\mathbb{N}$ , $\mathbb{Q}$ , $\mathbb{R}$ . Everybody knows how to construct $\mathbb{Q}$ and $\mathbb{R}$ from $\mathbb{N}$ . The definition of $\mathbb{N}$ needs some discussion.

$\mathbb{N}$ is the "least limit ordinal". Recall that an ordinal is a transitive set (i.e. such that $\forall y \forall z\ (((y \in x) \wedge (z \in y)) \Rightarrow (z \in x))$ whose elements are all transitive ; any two ordinals $x$ and $y$ satisfy the statement $(x \in y) \vee (x = y) \vee (y \in x)$ , which yields a total order between ordinals.

We put $sx = x \cup \{x\}$ (read "successor") and call limit ordinal an ordinal $x$ such that $(x \neq \emptyset) \wedge \forall y\, (x \neq sy)$ ; there is an axiom in Z.F. which implies the existence of an unique minimal limit ordinal ; we call it $\mathbb{N}$ in the corresponding extension of Z.F.

Each element of $\mathbb{N}$ is an ordinal ; we call it finite: Note that, a priori, this finiteness has nothing to do with "finite sets", that is sets which are not in one-to-one correspondance with a proper subset ; we need the axiom of choice to make both concepts equivalent.

The set $\mathbb{N}$ is totally ordered by $\in$ ; it satisfies the Peano axioms, which places arithmetics in the set theoretic framework.

9) Now, to what extend is our Z.F. more powerful than combinatorial mathematics ? As a first answer, we notice that everything in Z.F. is a combinatorial object... As a second answer, we recall that within Z.F., "actual infinity", real numbers, and a lot of non-finitary concepts are formalized !

To make these answers compatible, we must either extend the reach of finitary constructions (but J.W. Brouwer's work shows that this is illusive), either consider that there is a swindle somewhere... The situation is the same as for banknotes which replace bar gold and keep their value even when the coffers are empty !

Here we have to believe that Z.F. contains no contradiction (i.e. a theorem of type $A \wedge \neg A$ ), in other words is consistent ; note that the truth of finitary statements is warranted by the stability of our intuitive perceptions.

Thus we made a nice mechanic doll, like the old Dr Coppelius and get convinced that she is a living girl, for she satisfies so perfectly our mathematical wishes... But we get anxious if we consider theorem "$A \wedge \neg A \Longrightarrow B$" where B is an arbitrary formula : if there is one contradiction, the whole tree is rotten, because B may be replaced as well by $\neg B$ .

To avoid a complete catastrophe, we could restrict the reach of logical rules (as physisists do implicitely, for they only accept logical consequences of their principles which are not in contradiction with experimental results), or restrict the axioms and therefore loose some concepts we had to formalize.

Thus, we must prove by finitary induction that $0 = 1$ is not a theorem in Z.F. Unfortunately, Kurt Gödel proved in 1935 that this is not possible, for the infiniteness concept cannot be formalized completely without contradiction... Now, if you find a contradiction in your every-day-math, probably you made a mistake ; if not, you probably will have to face a crew of angry mathematicians next morning...

10) Extensions by definition are particular cases of conservative extensions, i.e. in which a formula of the old language is a theorem if and only if it is one in the old theory.

More generally, an extension of Z.F. introduces eventually new signs in the language with grammatical rules that extend the old ones, and additional axioms in the theory.

We accept the consistency of Z.F. without proof, but certainly not the <u>relative consistency</u> of its eventual extensions ; in other words, we must prove that if Z.F. is consistent, so is the extension.

Sometimes the negation of some new axiom is as consistent with Z.F. as this axiom itself ; we call it <u>independent</u> of the axioms of Z.F. (so is the axiom of choice - see further on).

11) From now on, we call "Z.F." any extension by definition of the initial theory ; its language contains at least names for the usual sets $\mathbb{N}$ , Q , $\mathbb{R}$ , Hilbert space, etc... which occur in mathematics.

## Lesson 2

## TO BE NATURAL OR NOT TO BE

Problem. Is every element of $\mathbb{N}$ natural ?

Comments. 1) In lesson 1, we considered the potential collection of natural integers, that is integers which are used in our everyday computations for practical purposes. Strictly speaking, they are the only numbers whose "existence" is well-formed. Some of them are perceptible entities, while the potential ones are suggested by the construction rule.

Moreover, our question is meaningless for an intuitionist - he never heard about a set $\mathbb{N}$ ...

On the other hand, a formalist would try to avoid the problem : he calls "natural integers" the elements of his set $\mathbb{N}$ ; here the word "natural" is only an allusion to the intuitive numbers used in preformalistic times to describe natural collections. Nowadays, there is an axiomatic set $\mathbb{N}$ , whose properties are close to those of the old naturals ; we are liberated from intuition...

- Well ! this is nice. But how do you number the pages of your book ? How do you count your money in the grocer's shop ? With old naturals or with elements of $\mathbb{N}$ ?

- No problem ! for each natural, we have a corresponding ordinal in $\mathbb{N}$ : starting with $\emptyset$ , translate your old instruction "put a bar behind" into "take the successor $sx = x \cup \{x\}$ " and you get it. For instance, to $0$ corresponds $\emptyset$ , to $1$ corresponds $\{\emptyset\}$ , to $2$ , $\{\emptyset , \{\emptyset\}\}$ , etc... ; so I may consider old naturals as elements of $\mathbb{N}$ and give them the same name as before.

- Why not ? Thus let us call these "effective" successors of $\emptyset$ natural integers. Now, is every element of $\mathbb{N}$ such an integer ? This is a metamathematical question ; before trying to answer, let us list some properties of natural integers, which may easily be proved by finitary induction.

2) The integers $0, 1, 2, 1000, 10^{10}, 10^{10^{10}}$ are natural ; sums and products of naturals are natural ; if $n$ is natural, so is every $m \leq n$ , and there is a prime natural $n' \geq n$ ; if $\omega \in \mathbb{N}$ is not natural and if $n$ is , then $\omega > n$ ; moreover there is a prime non-natural $\omega' > \omega$ .

Consider the famous problem of Fermat : to find integers $x, y, z$, $n > 2$ such that $x^n + y^n = z^n$ . Define $a \in \mathbb{N}$ by the following property : $a = 0$ if Fermat's problem has no solution and $a = \inf\{(x^2 + y^2 + z^2 + n^2), x^n + y^n = z^n\}$ if there is one.

For the time being, nobody is able to prove that $a$ is natural (or not), even if $a \neq 0$ .

Now, let us hear some formalists about our question.

3) First formalist : "This is a mystification. Consider the set $E$ of natural integers ; it contains $0$ , and is stable under succession ; hence, by Peano's axioms, $E = \mathbb{N}$ . Thus every integer is natural, what everybody knows !"

Second formalist : "Are you sure that $E$ is a set ? It's not clear for me, and I don't agree with your proof. When in doubt, refrain ! Therefore, I never use the "set" $E$ in a mathematical proof, but I may use natural integers in the grocer's shop..."

Third formalist : "I think we should try to formalize the question. We want to prove the statement "$\forall x \in \mathbb{N}$ , $x$ is natural" ; in other words "$\forall x \in \mathbb{N}$ , $x = 0$ or $x = 1$ or $x = 2$ , etc..." . There is a trouble here : little dots are not allowed in our formal language, nor infinitely long sentences ! It seems quite impossible to formalize the property "natural". I refrain too, this is diabolic...

A courageous formalist : "Why should'nt we try to formalize the opposite statement "There is a non natural $x$ in $\mathbb{N}$ ". We could introduce a new constant $\omega$ in our language and put near the old axioms a new scheme $\omega > 0$ , $\omega > 1$ , ... Here the little dots mean : The potential collection of axioms $\omega > i$ , where $i$ is any natural integer ; this is allowed, for in a deduction, you may only write on the paper an intuitively finite number of axioms of the scheme. Of course, this would be an extension of Z.F. and maybe lead to some contradiction or overcome the power of Z.F. It's possible that some old statement which cannot be proved within Z.F. has a proof by means of $\omega$".

First, second and third formalist : "This is dangerous. Stop playing with fire : you want to change mathematics ! "

An independent observer : "What anodd trouble ! You formalize such an intuitive concept as "the potentially infinite collection of natural integers" and then you get afraid that a little extension of your formalism could mean that your $\mathbb{N}$ is more powerful than you hoped for...

What would really be a nuisance for me is that infinity could be formalized with complete adequacy. The old dream of putting infinity in a formula is definitely lost since SKOLEM and GÖDEL's works... "

---

## Lesson 3

## A NON-STANDARD EXTENSION OF Z.F.

Metatheorem. <u>There is a conservative extension of Z.F. in which $\mathbb{N}$ contains non natural integers.</u>

<u>Comments</u> : 1) Following the courageous formalist in lesson 2, introduce a non defined constant $\omega$ in $\mathcal{L}_{Z.F.}$ and you get a language $\mathcal{L}_{Z.F.F.}$ . It contains "old" formulas (those of $\mathcal{L}_{Z.F.}$) and formulas with an occurence of $\omega$ . Define by finitary induction the axiom scheme $0<\omega$ , $1<\omega$ , ... , $i<\omega$ , ... where $i$ is arbitrary natural integer (considered as an element of $\mathbb{N}$ ) ; together with the axioms of Z.F. This yields a new theory Z.F.F.

Any old statement which is a theorem in Z.F. is also a theorem in Z.F.F.

We have to prove the converse, that is conservativeness of Z.F.F. with respect to Z.F. This is nearly obvious.

2) Let A be an old formula, which is a theorem in Z.F.F. Its proof needs at most a finite number of axioms "$i < \omega$" . If we replace in the formal deduction every occurence of $\omega$ by a large enough natural integer, we get a proof of A within Z.F. This ends the finitary proof of conservativeness : we described precise instructions to replace one proof by another.

3) Now the door is open : you may as well prove old statements within Z.F.F. There is an analogy with the use of complex numbers as a tool to solve real equations. But complex numbers can be defined from reals, while $\omega$ cannot be defined from the integers in Z.F.

Notice that undefined entities are used in mathematics a long time ago (maximal torii in Lie algebras, basis in vector spaces, maximal solutions in differential equations, ...), in relation with the choice axiom.

To justify Z.F.F., we didn't need this axiom, which is a stronger assumption on infinity than Peano's axioms. As a consequence, Z.F.F. is not very powerful ; we only introduce it to make the reader familiar with conservative extensions involving infinitesimals.

Further, we shall consider a conservative extension of Z.F.C. (Z.F. with axiom of choice) and use it as a practical frame for non-standard analysis.

4) Within Z.F.F., we have infinitely large elements in $\mathbb{N}$ (i.e. larger than each natural) ; they are "ideal elements", following an old vocabulary and this "idealisation" is the gain of Z.F.F. with respect to Z.F. Note that if $x$ is infinitely large, so is $x \pm i$ for every natural integer $i$ ; hence, there is no first infinitely large $x$ , nor last natural one. This proves that there is no subset $E$ of $\mathbb{N}$ containing only the naturals, nos subset containing only the non-naturals, for $E$ would be bounded (by $\omega$ ) without last element, which is a contradiction.

5) In relation with lesson I.8, consider a prime integer $q > \omega$ ; then the coset ring $\mathbb{Z}/q\mathbb{Z}$ is a field and the congruence class $\bar{i}$ of any natural integer $i$ has an inverse. Thus we have "natural rationals" in the field $\mathbb{Z}/q\mathbb{Z}$ ; it is easy to define a total order on them. Of course, this is not $\mathbb{Q}$ ...

6) As an illustration of Z.F.F., recall the paraphrase of Darwinism mentionned in the preface of this book. There are successive monkeys $0$ , $1$ , $2$ , ... ; there is a man $\omega$ (also $\omega - 1$ , $\omega - 2$ , ... , $\omega + 1, \omega + 2$ , ...) ; men and monkeys have the same properties (Peano axioms) and there is no collective criterion to distinguish them (no set of naturals). There is no last monkey and no first man...

Charles KONRAD certainly would not reject this analogy ! Other comparisons, for instance the concept of "observable" in physics, suggest that the mathematical art may sometimes find in experimental sciences a new look over itself !

7) Any conservative extension $T'$ of a theory $T$ may be used without restriction to prove statements of $T$ ; the information contained in both, as regards what is expressible in the language of $T$ , is strictly equivalent.

However, the proofs within $T'$ may be shorter, thanks to the use of ideal objects as intermedia. Moreover, the language of $T'$ may allow an easier description of intuitive concepts – "... plus conforme à l'art d'inventer" as Leibnitz said.

It's the same when you have to cross a river near its mouth : either you walk ashore until you can cross near the source, and go back on the other side, either you take a bridge, if there is one, near the mouth: This bridge joins the same places as the long way (this is conservativeness !), but you spare time and energy. If you join an island to the continent, you get an illustration for non conservative extensions : without a bridge, there is no path at all between both...

Non standard analysis is a way to throw bridges over some deep mathematical rivers... Note that intuitive proofs correspond to swimming in the river ; sometimes you succeed and reach the opposite bank, but it is not sure...

# Lesson 4

## LOOKING FOR ENLARGEMENTS IN Z.F.

THEOREM. Every set has an extension with partial transfer and weak idealization properties.

Comments. 0) In Section I, we based an infinitesimal calculus on the mysterious concept of enlargement, the properties of which were discribed in an "order of procedure".
In lesson 3, we outlined an axiomatic way to introduce infinitesimals. Both approaches are closely related, as we shall see later.
Now we explain in detail what enlargements are and how they arise as byproducts of set theory. At first, we try to find them in Z.F., with only a partial success as mentionned in the statement above (which is a theorem in Z.F., of course).

1) In the sequel, we have to do with finite sets : this means sets which are in one-to-one correspondance with finite ordinals, i.e. elements of $\mathbb{N}$ (see lesson II.1).

2) Recall that an n-ary relation on a set $E$ may be defined as a subset of $E^n$, the set of all mappings of the ordinal $n$ into $E$. Usually, this subset is called the graph of the relation, but we consider it as the relation itself.

The first order structure $\Sigma(E)$ based on $E$ is the union of $E$ with the set $\bigcup_{\substack{n\in\mathbb{N}\\ n\neq 0}} \mathcal{P}(E^n)$ of all relations on $E$.

An extension of $\Sigma(E)$ based on a set $\bar{E}$ is a mapping $\tau : \Sigma(E) \longrightarrow \Sigma(\bar{E})$ such that :

i) $x \in E \Longrightarrow \tau(x) \in \bar{E}$

ii) $A \in \mathcal{P}(E^n) \Longrightarrow \tau(A) \in \mathcal{P}(\overline{E}^n)$

iii) $\tau(\emptyset) = \emptyset$ and $\tau(E) = \overline{E}$

iv) $\tau$(diagonal of $E^n$) = diagonal of $\overline{E}^n$

v) $(x_1, \ldots, x_n) \in A \Longrightarrow (\tau(x_1), \ldots, \tau(x_n)) \in \tau(A)$ , where $x, x_i$, $A$ take all possible values.

A first order formula based on $E$ is a formula $\Phi$ of $\mathcal{L}_{Z.F.}$ , whose constants are elements of $\Sigma(E)$ and whose variables are specified to run on $E$ (that is, if $x$ occurs in $\Phi$ , we must find " $x \in E$ " at a good place in the formula). Such a formula reflects a property about some relations on $E$ and some elements of $E$ ; quantifications only concern variables running on $E$ (and not on $\mathcal{P}(E)$ , $\mathcal{P}(\mathcal{P}(E))$ , ... , which is the case for higher order formulas).

Replacing in $\Phi$ each constant by its image under some extension $\tau$ , we get a first order formula $\overline{\Phi}$ based on $\overline{E}$ , called the transfered form of $\Phi$ via $\tau$ .

For instance, " $\forall x \forall y ((x \in \mathbb{N} \wedge y \in \mathbb{N}) \Longrightarrow x + y = y + x)$ " is a first order formula based on $\mathbb{N}$ : Its transfered form in an extension of $\Sigma(\mathbb{N})$ is

$$" \forall x \forall y ((x \in \overline{\mathbb{N}} \wedge y \in \overline{\mathbb{N}}) \Longrightarrow x \oplus y = y \oplus x) "$$

where $\oplus$ is a name for the transfered form of $+$ . (Notice that the variables are unchanged and that $=$ is again the equality.) On the other hand, the sentence "every subset of $\mathbb{N}$ has a first element" cannot be formalized as a first order formula based on $\mathbb{N}$ . But, following lesson I.8, we may write it down as a first order formula based on $\mathbb{N} \cup \mathcal{P}(\mathbb{N})$ , reminding that $\tau(\mathcal{P}(\mathbb{N}))$ is not the same as $\mathcal{P}(\tau(\mathbb{N}))$ .

Now, we say that an extension $\tau$ has the transfer property if, for every closed first order formula $\Phi$ based on $E$ , $\Phi \Longleftrightarrow \overline{\Phi}$ is a theorem in Z.F.

To find criteria for the transfer property, notice that any first order formula $\Phi$ on $E$ may be expressed by means of intersections, complements and projections within $\Sigma(E)$ (projections are mappings $\pi_i : E^n \longrightarrow E^{n-1}$ of type $(x_1, \ldots, x_n) \longrightarrow (x_1, \ldots, x_{i-1}, x_{i+1}, \ldots, x_n)$ ) ; this is due to the fact that $\wedge$, $\neg$ and $\exists x$ are the only logical operations on formulas used to construct $\mathcal{L}_{Z.F.}$

From this it is easy to check that $\tau$ <u>has the transfer property if and only if for every subset</u> A <u>and</u> B <u>of</u> $E^n$ , $n \in \mathbb{N}$ , <u>one has</u>

vi) $\tau(A \cap B) = \tau(A) \cap \tau(B)$ (transfer of $\wedge$ )

vii) $\tau(E^n - A) = \overline{E}^n - \tau(A)$ (transfer of $\neg$)

viii) $\tau(\pi_i(A)) = \pi_i(\tau(A))$ (transfer of $\exists x$ ).

As a consequence, note that $\tau$ is one-to-one into (transfer the diagonals by iv) and apply vii) ).

3) Consider a binary relation $\rho \in \mathcal{P}(E \times E)$ and assume that <u>for every finite subset</u> F <u>of its source</u> $\pi_2(\rho)$ , <u>there exists</u> $y \in E$ <u>such that, for every</u> $x \in F$ , $(x, y) \in \rho$ . Then we say that $\rho$ is <u>idealizable</u> ("concurrent" in [R]).

Call $\Gamma_2(E)$ the set of such relations on E (this is "type $\Gamma_2$ " in section I). The basic relations of analysis and topology, namely the order relation in $\mathbb{R}$ and the inclusion of open sets with a common point in topology are idealizable ; this is the reason why Non Standard Analysis works mainly in these areas...

If $\rho \in \Gamma_2(E)$ , usually there is no $y \in E$ related to every $x \in \pi_2(\rho)$ (e.g. no minimal open neighbourhood of a point in a topological space, or no highest real number). But in an extension $\tau$ of $\Sigma(E)$ , it may exist an $\omega \in \overline{E}$ such that $(\tau(x), \omega) \in \tau(\rho)$ for every $x \in \pi_2(\rho)$ . Such an $\omega$ is called an <u>ideal element for</u> $\rho$ .

For instance, consider the trivial relation $\rho = E \times E$ . It is clearly idealizable and every element of E is ideal in the trivial extension $\tau$ = identity. On the other hand, the relation $\neq$ , whose graph is the complement of the diagonal in $E \times E$ , is idealizable if and only if E is infinite ; it has no ideal element in the trivial extension.

Any idealizable relation without ideal element in the trivial extension, which has nothing to do with analysis or topology is the foundation of some other branch of mathematics. This is an evidence, but it seems hard to device an exam-

ple. The main idealizable relations we know are strongly idealizable in the following sense : there is a mapping $F \longrightarrow y(F)$ which associates to every finite subset $F$ of $\pi_2(\rho)$ an $y(F)$ such that, for every $x \in F$, $(x, y(F)) \in \rho$. (For instance $y(F) = \text{Sup } F$ in $\mathbb{R}$, or the intersection of a finite family of open neighbourhoods in topology.)
An exception is the relation $\neq$, for which we cannot prove within Z.F. that it is strongly idealizable on any set $E$. Let $\gamma_2(E) \subset \Gamma_2(E)$ be a set of all strongly idealizable relations on $E$.
An extension $\tau$ of $\Sigma(E)$ has the idealization property if every $\rho \in \Gamma_2(E)$ has an ideal element in $E$.

Replace $\Gamma$ by $\gamma$ and you get the weak idealization property.
Finally, an extension $\tau$ of $\Sigma(E)$ is called an enlargement of $E$ (resp. weak enlargement) if it has the transfer property and the idealization (resp. weak idealization) property.

Note that in section I, we "put stars" in place of $\tau$.

Exercise. Prove that $E$ is finite if and only if the trivial extension is an enlargement. What about other enlargements of $E$ ?

5) Our aim is to find as much enlargements as possible, in order to use ideal elements as intermedia in everyday practice. Classical extensions, e.g. from $\mathbb{N}$ to $\mathbb{C}$, use set theoretic constructions (such as products) to get bigger sets and quotients to reduce them if necessary. Let us try to construct enlargements of a set $E$ in the same spirit.
Choose an index set $I$ and consider $E^I$. For each $\alpha \in E$, call $\tau(\alpha)$ the constant mapping $i \longrightarrow \alpha$ (i.e. $\tau(\alpha) = (\alpha, \alpha, \dots, \alpha)$ ) and for each $A \in \mathcal{P}(E^n)$, put $\tau(A) = \{(x^1, \dots, x^n) \in (E^I)^n, \ (x_i^1, \dots, x_i^n) \in A \text{ for every } i \in I\}$. It is easy to verify that $\tau$ is an extension of $\Sigma(E)$ based on $E^I$. This "generalized trivial extension" (for $I = \{\text{one point}\}$ you get the trivial one) agrees with intersections, inclusions and projections. Hence it has the transfer property for $\wedge$, $\Longrightarrow$ and $\exists x$, $\neq$, but negation, universal quantification $\forall x$ and $\vee$

don't transfer if $I$ has more than one element. We say that $\tau$ has only the <u>partial transfer property</u>. Note that the only idealizable relations on $E$ which have an ideal element in $E^I$ are those which have one in $E$. Thus, for the time being we gain little. Ideal elements should appear in some quotient extension (recall Cauchy sequences to get $\mathbb{R}$ from $\mathbb{Q}$ ...).

6) To this end consider a family of subsets of $I$, that is an element $\mathcal{F} \in \mathcal{P}(\mathcal{P}(I))$ ; define a binary relation $\theta$ on $E^I$ by

$$(x, y) \in \theta \iff \{i , x_i = y_i\} \in \mathcal{F} .$$

Then $\theta$ is an equivalence relation if and only if $\mathcal{F}$ has properties

$(f_1)$ $I \in \mathcal{F}$ ( $\theta$ is reflexive)

$(f_2)$ $(U \in \mathcal{F}) \wedge (V \in \mathcal{F}) \Longrightarrow (U \cap V \in \mathcal{F})$ ( $\theta$ is transitive).

Call $\overline{E} = E^I/\mathcal{F}$ the quotient set and $q : E^I \longrightarrow \overline{E}$ the natural projection.
For every $A \in \mathcal{P}(E^n)$, let $\overline{A}$ be the set of all $(x_1, \ldots, x^n) \in (E^I)^n$ such that $\{i \in I , (x_i^1, \ldots, x_i^n) \in A\} \in \mathcal{F}$.
Then $\overline{A}$, considered as a relation on $E^I$, is compatible with $\theta$ if and only if $\mathcal{F}$ has property

$(f_3)$ $(U \in \mathcal{F}) \wedge (V \supset U) \Longrightarrow V \in \mathcal{F}$ .

Thus, from $(f_1)$ $(f_2)$ $(f_3)$ we get an extension $\overline{\tau}$ of $\Sigma(E)$ based on $\overline{E}$, with $\overline{\tau} = q \circ \tau$ on $E$ and $\overline{\tau}(A) = \{(q(x^1), \ldots, q(x^n)), (x^1, \ldots, x^n) \in \overline{A}\}$ for $A \in \mathcal{P}(E^n)$. This extension has the transfer property for $\wedge$, $\Longrightarrow$, $\exists$.
We list some additional properties of $\mathcal{F}$, which translate into properties of $\overline{\tau}$.

$(f_4)$ $\emptyset \notin \mathcal{F}$ ( $\neq$ transfers, hence $\overline{\tau}$ is one to one into)

$(f_5)$ $U \in \mathcal{P}(I)$, $U \neq \emptyset \Longrightarrow (U \in \mathcal{F}) \vee ((I - U) \in \mathcal{F})$ ( $\neg$ transfers)

$(f_6)$ $\bigcap_{U \in \mathcal{F}} U = \emptyset$ (in case $E$ is infinite, $\neq$ has an ideal element in $\overline{E}$, hence $\overline{\tau}$ is not onto).

The proofs are straightforward.
Now some vocabulary. An $\mathcal{F}$ with properties $(f_1)$ $(f_2)$ $(f_3)$ $(f_4)$ is called a <u>filter on</u> $I$ : it retains some subsets of $I$ and leaves the others "pass through

it". Two elements of $E^I$ are $\theta$-related if the subset of $I$ on which both coincide is retained by the filter $\mathcal{F}$. We call $\bar{E}$ <u>a filtered power of</u> $E$.
For instance $\mathcal{F} = \{I\}$ is a filter for which $\bar{E} = E^I$.
A filter with $(f_4)$ and $(f_5)$ is called an <u>ultrafilter on</u> $I$.
A filter with $(f_6)$ is called <u>free</u>, for the subsets it retains have no common element.
Note that the rough filter $\mathcal{F} = \{I\}$ is nor free, nor ultra. Let us summarize this discussion as following :

" $\mathcal{F}$ <u>is a filter on</u> I " $\Longleftrightarrow$ " $\tau$ <u>is a well defined extension with partial transfer property</u>"

" $\mathcal{F}$ <u>is an ultrafilter</u>" $\Longleftrightarrow$ " $\bar{\tau}$ <u>has transfer property</u>"

" $\mathcal{F}$ <u>is a free ultrafilter</u>" $\Longleftrightarrow$ " $\bar{\tau}$ <u>has transfer property and, if</u> $E$ <u>is infinite,</u> $\neq$ <u>has an ideal element in</u> $\bar{E}$ ".

7) The set of filters on an index set $I$ is partially ordered as follows : $\mathcal{F}'$ is <u>finer</u> than $\mathcal{F}$ if $U \in \mathcal{F}$ implies $U \in \mathcal{F}'$. Thus $\mathcal{F} = \{I\}$ is the roughest filter on $I$. One of the first historical result on filters was the following characterization of ultrafilters.

<u>An ultrafilter is a maximal element for the fineness relation.</u>

<u>Proof</u>. If $\mathcal{F}'$ is finer than an ultrafilter $\mathcal{F}$, every $U \in \mathcal{F}'$ is also in $\mathcal{F}$, otherwise $I - U \in \mathcal{F}$, hence $I - U \in \mathcal{F}'$ which contradicts $(f_2)$ $(f_4)$ $(f_5)$. Thus $\mathcal{F}' = \mathcal{F}$.
Conversely, if $\mathcal{F}$ is maximal and if $U \in \mathcal{P}(I)$, $U \neq \emptyset$, $U \notin \mathcal{F}$, adjoin to $\mathcal{F}$ all subsets of $I$ which contain $I - U$. You get a filter $\mathcal{F}'$, which is finer than $\mathcal{F}$. As $\mathcal{F}' = \mathcal{F}$, infer that $I - U \in \mathcal{F}$.
As an example of ultrafilter, consider a point $a \in I$ and $\mathcal{F} = \{U \subset I\ ,\ a \in U\}$. This ultrafilter is not free, of course. Indeed, property $(f_3)$ implies that every ultrafilter which is not free is of this type.

<u>Exercise.</u> Try to find a free ultrafilter within Z.F. (begin with $I$ finite). If you don't succeed, see lesson 5.

8) To get an extension $\overline{T}$ with at least weak idealization property (recall that $\neq$ need not have an ideal element, for it is not a strongly idealizable relation in general), it is natural to seek a convenient filter on the set I of all finite subsets of E.

If $\rho \in \gamma_2(E)$, we have a mapping f which associates to each finite subset F of the source $S(\rho)$ an element $f(F) \in E$ such that $(\alpha, f(F)) \in \rho$ for every $\alpha \in \mathcal{F}$. Define $y : I \longrightarrow E$ as follows :

$$\begin{cases} \text{if } i \cap S(\rho) = \phi , & y_i \text{ is arbitrary.} \\ \text{if } i \cap S(\rho) \neq \phi , & y_i = f(i \cap S(\rho)) . \end{cases}$$

Let $\mathcal{F}$ be a filter on I. In order that the equivalence class of y in $E^I/\mathcal{F}$ be an ideal element for $\rho$, a necessary and sufficient condition is that, for every $u \in S(\rho)$, the set $F^u_\rho = \{j \in I , (u, y_j) \in \rho\}$ (which contains $\{j \in I , u \in j \cap S(\rho)\}$ ) is an element of $\mathcal{F}$. This suggests to construct a filter with these $F^u_\rho$, $\rho$ running on $\gamma_2(E)$ and u on $S(\rho)$.

Indeed, the intersection of any finite family $\{\mathcal{F}^{u_k}_{\rho_k}\}_{1 \le k \le n}$ contains $j = \{u_1, \ldots, u_n\}$. These intersections set up a filter basis, that is condition " $U \in \mathcal{F} \Longleftrightarrow U$ contains some finite intersection of $F^u_\rho$ " yields a filter $\mathcal{F}$. Moreover, for this filter and also for any finer one, we have an extension of $\Sigma(E)$ with partial transfer and weak idealization properties. Note that in case $\neq$ is strongly idealizable, the filter $\mathcal{F}$ is free (e.g. $E = \mathbb{R}$ ) ; thus, if we want a weak enlargement of E, we have to find an ultrafilter that is finer than a free filter. Such an ultrafilter is itself free and it is high time to get an answer about our exercise above.

9) The use of extensions with partial transfer and weak idealization properties to do some non standard mathematics is not very interesting. For instance, in $\mathbb{R}$, as < is strongly idealizable, we get infinitesimals ; direct parts of non-standard characterizations are true, e.g. f continuous implies " $x \sim x_0 \Longrightarrow \overline{f}(x) \sim \overline{f}(x_0)$ ", but not their converses because the negation doesn't transfer.

# Lesson 5

## WEAK ENLARGEMENTS AND ULTRAFILTERS

THEOREM. The following statements are equivalent :

i) Every set E has a weak enlargement ;

ii) For every filter $\mathcal{F}$ on a set K, there exists an ultrafilter $\mathcal{U}$ on K which is finer than $\mathcal{F}$.

Comments. 0) This is a theorem in Z.F. But i) or ii) may not be theorems ! We discuss this point later.

1) To prove ii) $\Longrightarrow$ i), consider the filter $\mathcal{F}$ in § 8, lesson 4 ; from ii) we get a finer ultrafilter $\mathcal{U}$ and thus a weak enlargement, for the corresponding extension has the whole transfer property.
Usually $E^I/\mathcal{U}$ is called an ultrapower of E.

2) To prove i) $\Longrightarrow$ ii), put $E = K \cup \mathcal{P}(K) \cup \mathcal{P}(\mathcal{P}(K))$ and consider the following binary relation $\rho$ on E.

" $(A, \Phi) \in \rho$" $\Longleftrightarrow$ " $A \in \mathcal{P}(K)$, $\Phi \in \mathcal{P}(\mathcal{P}(K))$ and $\Phi$ is a finer filter than $\mathcal{F}$ such that $A \in \Phi$ or $K - A \in \Phi$ ".

This relation is strongly idealizable ; indeed, if $\{A_1, \dots, A_n\}$ is a finite subset of the source $\mathcal{P}(K)$, we construct a filter, finer than $\mathcal{F}$, containing each $A_i$ or its complement, by means of the following procedure used n times, beginning with $\mathcal{F}$ :

- if $A \in \mathcal{P}(K)$ intersects all elements of a filter $\Phi$, adjoin it to $\Phi$ with all $B \in \mathcal{P}(K)$ such that $A \subset B$. This gives a finer filter.

- if for some $U \in \Phi$, $A \cap U = \emptyset$, then $K - A \supset U$ ; hence $K - A \in \Phi$ and we keep $\Phi$ as finer filter.

Let $\overline{E}$ be the basis of some weak enlargement of E and $\overline{\in}$ the transfered form of $\in$ , which is considered as a binary relation on $E$ . Relation $\rho$ has an ideal element $\Omega$ in $\overline{E}$ , that is an element of $\overline{\mathcal{P}(\mathcal{P}(K))}$ which satisfies the transfered form of "filter, finer than $\mathcal{F}$ " and is such that, for each $A \in \mathcal{P}(K)$ , one has $\overline{A} \,\overline{\in}\, \Omega$ or $\overline{K} - \overline{A} \,\overline{\in}\, \Omega$ .

Put $\mathcal{U} = \{B \in \mathcal{P}(K) \ , \ \overline{B} \,\overline{\in}\, \Omega\}$ ; using transfer property, it is easy to check that $\mathcal{U}$ is an ultrafilter, which is finer than $\mathcal{F}$ .

3) Now we have to discuss ii). Is it a theorem in Z.F. or not ? Note first that on a finite set $K$ there is no free ultrafilter : it would contain no singletons, hence all subsets with $n-1$ elements ; by repeated intersections, we get singletons...

Now start with a free filter on $K$ ; there is no finer ultrafilter for it would also be free. Conclude : ii) is not a theorem in Z.F.

" - Sorry ! You have to provide an example of a free filter on a finite set.

- ???...

- I think that there is none, indeed ! the set of all filters on $K$ is finite and ordered ; in such a set every element has a maximal successor, that is an ultrafilter in our case ; so starting with a free filter you get a free ultrafilter... Conclude : ii) may be a theorem in Z.F.

- On infinite sets, there is no problem. We have examples of free filters (e.g. the very complicated filter in § 8, lesson 4, whenever $\neq$ is strongly idealizable). But at the time being, nobody is able to produce a free ultrafilter in Z.F. ... Furthermore, logicians may find good arguments to prove the relative consistency of " $\neg$ (ii) " with Z.F. ... "

So far, we have a theory Z.F.U., that is Z.F. with ii) as a supplementary axiom, also Z.F.e. (i.e. Z.F. with ii) ) which is the same theory as Z:F.U., and Z.F.($\neg$ U) with $\neg$ (ii) as an axiom. These theories are non conservative extensions of Z.F. If we consider Z.F.($\neg$U) as a basis for our mathematics, we have not enough enlargements to do Non-Standard-Analysis. This is a good reason to prefer Z.F.U., although we are not sure of its relative consistency with Z.F.

But there is a more important reason : in the next lesson, we shall see that axiom ii) is a consequence of the axiom of choice, which is commonly accepted in mathematical foundations.

4) Exercise. Prove the theorem of ARTIN-SCHREIER within Z.F.e. : a commutative field $K$ may be ordered iff every finite sum of non zero squares is non zero.

Hint. If $K$ is ordered (as a field), and if $x_i \neq 0$ , then $x_i^2 > 0$ and $\Sigma\, x_i^2 > 0$ . Conversely, use the following condition to get an order in $K$ : there is a subset $P$ (think at the positive elements) such that

$$(1)\quad \begin{cases} P+P \subset P \\ P \cdot P \subset P \\ P \cap (-P) = \{0\} \\ P \subset \Sigma\, K^2 \end{cases} \qquad \text{and} \quad (2)\quad P \cup \{-P\} = K .$$

If $A \subset K$ satisfies (1) and if $x \notin A \cup (-A)$ , then $A - xA$ satisfies (1) (little computation). Now, consider a finite subset $\{x_1 , \ldots , x_n\}$ of $K$ . The inductive relations

$$A_o = \Sigma\, K^2 \quad \text{and} \quad A_{i+1} = \begin{cases} A_i & \text{if } x_i \in A_i \cup (-A_i) \\ A_i - xA_i & \text{if } x_i \notin A_i \cup (-A_i) \end{cases}$$

define a subset $A_n$ which satisfies (1) and such that $\{x_1 , \ldots , x_n\} \subset A_n \cup (-A_n)$ . This proves that the binary relation $\in$ between elements of $K$ and subsets of $K$ satisfying (1) is strongly idealizable. In a weak enlargement of $K \cup \mathcal{P}(K)$ , its ideal element $\Omega$ yields a set $P = \{x \in K \ , \ x \,\bar{\in}\, \Omega\}$ satisfying (1) and (2).

Remark. Classical proofs use Zorn's lemma, which immediatly provides a maximal element $P$ satisfying (1) and (2). This is very short, of course.

But we aim to prove the theorem in a weaker theory than Z.F. with the axiom of choice (which is equivalent to Zorn's lemma), to get some familiarity with enlargements.

A lot of other questions where Zorn's lemma is the usual tool may be solved using weak enlargements. The reader may discuss this point about the following statements :

- every vector space has a basis.
- every Lie algebra has a maximal abelian subalgebra.
- the "Nullstellensatz" of Hilbert.
- the existence of maximal solutions for differential equations.

---

## Lesson 6

## ENLARGEMENTS IN Z.F.C.

THEOREM. (C) $\Longrightarrow$ (E) where (C) and (E) are the following statements :

(C) For every set $E$ and every family $\alpha$ of non-empty subsets of $E$, there exists a mapping $f : \alpha \longrightarrow E$ such that, for every $F \in \alpha$, $f(F) \in F$.

(E) Every set has an enlargement.

Comments. 0) This is a theorem in Z.F.

1) The mapping $f$ in (C) is a function of choice : it chooses in each $F \in \alpha$, with $\alpha \in \mathcal{P}(\mathcal{P}(E))$ an element $f(F)$. Within Z.F. such a choice is possible for any family of 1, 2, 3, etc... (natural numbers, as in lesson II.2.) elements ; but statement (C) - called axiom of choice - is a very strong assumption on infinity and the main motive of dissension among mathematicians in a recent past. Note that (C), implies Zorn's lemma :

For every ordered set, there exists a maximal totally ordered subset, i.e. which cannot be extended into a larger totally ordered subset.

2) Proof of the theorem. From (C) infer that every idealizable relation is strongly idealizable. Hence weak enlargements are enlargements. Now consider the set of all filters on $I$, which are finer than a given filter $\mathcal{F}$. By Zorn 's

lemma, there exists a maximal totally ordered subset of this set. The union of all filters of this subset is a filter which has no finer one (use maximality) ; this maximal filter is an ultrafilter which is finer than $\mathfrak{F}$ . By lesson 5, infer that every set has an enlargement (not unique, of course).

3) Call Z F C (resp. Z F E ) the non conservative extensions of Z.F. with (C) (resp. (E)) as a supplementary axiom. We have :

$$\begin{cases} ZFU = ZFe \\ ZFC \text{ contains } ZFE \\ ZFE \text{ contains } ZFe \end{cases}$$

Thus the relative consistency of Z F E (or Z F e ) is a byproduct of the relative consistency of Z F C . The last has been proved (with finitary arguments), but it is not an evidence, of course. (see for instance the book of J.R. Schoenfield "Mathematical logic" Addison Wesley 1967, or some other book including set theory.)

The main consequence is that if you agree with Z.F.C. as background of your everyday mathematics, you may use as well its byproduct Z.F.E. This time, you get something for nothing !

Subtle people may wonder about the difference between Z.F.E. and Z.F.C. Specialists can prove them that Z.F.C. is slightly stronger than Z.F.E., i.e. axiom (C) may be hard to prove in its full strength within Z.F.E. Also they can prove that ZF ($\neg$C) is relatively consistent with Z.F., as well as Z.F.C. But such an information as $\neg$C has no intuitive foundation and mathematics based on ZF ($\neg$C) would not be more powerfull than those based on Z F, in some sense. However this implies that (C) is not a theorem in Z.F.

4) An equivalent form of (C) is Zermelo's lemma (well ordering theorem) For every set E , there is a one-to-one mapping of E onto some ordinal : the least of such ordinals is called the cardinal number of E .

From this, it is easy to deduce the well-known characterization of finiteness within Z.F.C.

A set E is finite if and only if there is no one-to-one mapping of E onto some proper subset of E .

Axiom (C) implies the existence of objects which cannot be constructed with elementary set theoretic operations. The most surprising of them is certainly a model of Z.F.C.

It is a pair $(E, G)$ where $G$ is a subset of $E \times E$ such that, if we replace " $x = y$ " by " $(x, y) \in \text{diagonal}(E \times E)$ " , " $x \in y$ " by " $(x, y) \in G$ " in the formulas of $\mathcal{L}_{Z.F.}$ , and restrict each variable to run in $E$ , then all axioms (hence all theorems) of Z.F.C. give again theorems after these operations.

For instance, the theorem " $\forall x , x \notin x$ " is translated in the theorem " $\forall x \in E , (x, x) \notin G$ ".

In other words, the elements of $E$ play the part of "sets" and $G$ is a binary relation on $E$ which plays the part of "membership". To be convinced about the complexity of $(E, G)$ , recall that, apart from some restrictive ones, all axioms of ZFC insure some "existence" ; the company $E$ has to be numerous...

Be carefull about the canonical mistake : $E$ is not "the set of all sets", which is not allowed in Z.F. To avoid it, don't confuse " $x \in y$ " and " $(x, y) \in G$ " ; the second formula is only correct if you adjoin to it $x \in E$ and $y \in E$ .

Other theories may have models in Z.F.C. . after translating their language in $\mathcal{L}_{Z.F.}$ , their axioms must become theorems in Z.F.C. Such a theory is relatively consistent with Z.F. (a poor information, indeed...) because any contradiction would translate into a contradiction in Z.F.C., hence in Z.F. The converse is more interesting ; it is the key of model theory .

If a theory is consistent with Z.F., it has a model in Z.F.C. (completness theorem)

The proof is a complicated consequence of axiom (C).

Remark : To avoid some misunderstanding about models see the discussion lesson 7 § 14.

5) A lot of elementary theories - e.g. group, ring, field theories - have models even in Z.F. But we are interested in very big theories, like Z.F.C. itself or other extensions of Z.F. and axiom (C) is fundamental.

Let us give two examples :

- Let $\omega$ be a constant, defined by " $A(\omega)$ " in some extension by definition of Z.F.C. Consider a model $(E, \mathcal{G})$ of Z.F.C. ; to get a model of the extension, we have to find an $a \in E$ which plays the part of $\omega$ in such a way that $\widetilde{A}(a)$ is a theorem, where $\widetilde{A}$ is the translation of $A$ in the model.
Now the translation of theorem $\exists ! x \, A(x)$ yields such an $a$ . This proves, in a complicated way, that the extension is consistent. Clearly, the finitary proof is more informative !

- This is also the case for our theory Z.F.F. in lesson 3. The finitary proof of consistency was immediate. Let us give the model-theoretic one.
Consider a model $(E, \mathcal{G}, \bar{0}, \bar{1}, \ldots, \nu)$ of Z.F.C. with a "name" in $E$ for each natural integer and for $\mathbb{N}$ . Adjoin a (non defined) constant $\omega$ to the language and the axioms $0 < \omega$ , $1 < \omega$ , ... to get Z.F.F. Consider an enlargement $({}^*E, {}^*\mathcal{G}, {}^*\bar{0}, {}^*\bar{1}, \ldots, {}^*\nu)$ of the first model. By transfer, this is again a model of Z.F.C. (replace $=$ by the diagonal of ${}^*E \times {}^*E$ and $\in$ by ${}^*\mathcal{G}$ and $0, 1, \ldots, \mathbb{N}$ by ${}^*\bar{0}, {}^*\bar{1}, \ldots, {}^*\nu$ ) and we hope to find an $\bar{\omega} \in {}^*E$ such that $(\bar{\omega}, {}^*\nu) \in {}^*\mathcal{G}$ and $({}^*\bar{0}, \bar{\omega}) \in {}^*\mathcal{G}$ , $({}^*\bar{1}, \bar{\omega}) \in {}^*\mathcal{G}$ , etc...
To this end, note that the following relation $\rho(x, y)$ is idealizable :

" $x \in E \wedge y \in E \wedge (x, \nu) \in \mathcal{G} \wedge (y, \nu) \in \mathcal{G} \wedge (x, y) \in \mathcal{G}$ ".

Thus it has an ideal element $\bar{\omega} \in {}^*E$ such that $(\bar{\omega}, \nu) \in {}^*\mathcal{G}$ and $\forall x \in E$ , $((x, \nu) \in \mathcal{G} \Longrightarrow ({}^*x, \bar{\omega}) \in {}^*\mathcal{G})$ which is much more than expected.

6) It is clear that an enlargement ${}^*({}^*E)$ of an enlargement ${}^*E$ of $E$ is itself an enlargement of $E$ . More generally, consider a sequence $E_i$ such that $E_o = E$ and $E_{i+1}$ is an enlargement of $E_i$ ; its direct limit, whose basis is the quotient of $\bigcup_{i \in \mathbb{N}} E_i$ by identification of $x \in E_i$ with its image in $E_j$ whenever $i < j$ , is also an enlargement of $E$ .
Such a "limit enlargement" ${}^{\circledast}E$ keeps a trace of its internal constitution, called <u>idealization-with-parameters</u> property .
<u>Let</u> $\rho \subset E^{2+k}$ <u>be a</u> $(2+k)$-<u>ary relation such that</u>, <u>for each</u> $t_1, \ldots, t_k \in {}^{\circledast}E$

<u>and each finite subset</u> $F$ <u>of</u> $E$ <u>there is an element</u> $y \in {}^{*}E$ <u>such that</u> ${}^{*}\rho(x, y, t_1, \dots, t_k)$ <u>for every</u> $x \in F$. <u>Then, given</u> $t_1, \dots, t_k \in {}^{*}E$, <u>there exists an ideal element</u> $\xi \in {}^{*}E$ <u>such that</u> ${}^{*}\rho(x, \xi, t_1, \dots, t_k)$ <u>for every</u> $x \in E$.

The proof is left to the reader.
Note that for $k = 0$, condition "there is an element $y \in {}^{*}E$ such that ${}^{*}\rho(x, y)$ for every $x \in F$" is equivalent by transfer to "there is an element $y \in E$ such that $\rho(x, y)$ for every $x \in F$", because ${}^{*}F = F$ (with the natural identification of $E$ as a subset of ${}^{*}E$).
W.A.J. Luxemburg has given an example of an enlargement which has not this strong idealization property (see [L]).
From now on, we use only limit enlargements, since it is not more expensive...

7) In such an enlargement we have standard objects - those of $\Sigma(E)$ and internal ones - those of $\Sigma({}^{*}E)$ ; by identification, we may consider standard objects as internal and work as in section I.
In the next lesson, we shall try to describe in an axiomatic way the interplay between the "standard" and "internal" qualities.

# Lesson 7

# INTERNAL SET THEORY

Programme. Find a conservative extension of Z.F.C. in which the relations between standard and internal objects in limit enlargements are described as well as possible by the axiom system.

Comments. 0) In section I, we used N.S.A. under its original presentation, following A. Robinson, by means of enlargements. In section II, we related enlargements with ultrapowers - a well-known concept in classical mathematics.

From a didactical point of view, enlargements seem quite mysterious to non logically-minded people and such purely existential entities, moreover not uniquely characterized by their properties, may be rejected as "extramathematicals", as it was the case a few centuries ago for "imaginary numbers".

Of course, this attitude is rather irrationnal, since enlargements rely only on the axiom of choice, which is of common use.

Now, a conservative extension of Z.F.C. allowing non-standard tricks like infinitesimals without introducing mysterious entities would have at least the following advantages :

- to satisfy the preference of mathematicians for "rules" rather than for entities,
- to yield a formal argument for N.S.A.'s efficiency : some complicated proofs in ZFC may be shortened within the extension, due to a richer language which allows new intermediate formulations of old statements (recall the "bridges" in lesson II.3.),
- to clarify the following point : how is it possible to get a proof within ZFC from a non-standard proof ? Is there some mechanical trick to do this ?

1) Some authors (e.g. G. Kreisel [K]) proposed answers to the program-

me above, that A. Robinson clearly mentionned at the end of his book. These answers remained useless ; they seemed to be some exercises for logicians rather than a practical approach to N.S.A.

It was E. NELSON [N] who proposed an "Internal set theory" (I.S.T.) as an axiomatic description of N.S.A. intended for use in the every-day mathematical life. The inconvenience of such a theory is the artificial character of its axioms, for somebody who never heard about N.S.A. in its original form (nevertheless, elementary extensions as in lesson 3 may prepare the minds).

For this reason we started this book with enlargements, a very natural concept, as a preparation to non-standard ideas and in this section we end with I.S.T., proving its conservativeness by means of enlargements. In section III and IV, we use I.S.T. for sake of simplicity.

2) Consider a limit enlargement $({}^*E\, ,\, {}^*G)$ of some model $(E\, ,\, G)$ of Z.F.C. Complete the language $\mathcal{L}_{Z.F.}$ with a new sign " St " (read " x is standard" for " st x ") to be put before any constant or variable (this is called a <u>monadic predicate symbol</u>) ; in the new language $\mathcal{L}$ we may formulate all properties of $({}^*E\, ,\, {}^*G\, ,\, E)$ which are true for any model and any enlargement of it (of course, particular models may have other properties - e.g. models of a non conservative extension of Z.F.C. - but we don't consider them).

These properties are the following :

- by transfer, $({}^*E\, ,\, {}^*G)$ is a model of Z.F.C. (take ${}^*G$ as the graph of $\in$ and the diagonal of ${}^*E \times {}^*E$ as the graph of $=$ ).

- transfer and idealization (with parameters) properties between $E$ and ${}^*E$, whenever " St x " is replaced by $x \in E$ :

- consider an $x \in E$ and a formula $C(Z)$ of $\mathcal{L}$ .

Then there is a set $\{z \in E\, ,\, (z\, ,\, x) \in G \wedge C(z)\}$ , which is a subset of $E$ in the "external" theory Z.F.C. . Hence there is in the model a $y \in E$ which plays the part of "the standard subset of $x$ whose standard elements are precisely those standard elements of $x$ satisfying the property $C$ " (Note that the subset axiom of Z.F.C. only works for formulas of $\mathcal{L}_{Z.F.}$ , when applied to the model $({}^*E\, ,\, {}^*G)$.

If we take these properties as axioms for a theory T , we know a priori that T is consistent, since it has a model. But whether T is a conservative extension of Z.F.C. is not obvious ; some improvement of the model theoretical argument is necessary.
At this point, we are motivated for a survey of Nelson's paper.

3) First let us state precisely the axioms of I.S.T. as outlined above. Call internal the formulas of $\mathcal{L}_{Z.F.}$ in $\mathcal{L}$ , i.e. external formulas (those of $\mathcal{L}$ ) without any occurence of " St " in them. Following Nelson, we use the following abbreviations :

$\forall^{St} x\, A$ for $\forall x\ (St\ x \Longrightarrow A)$

$\exists^{St} x\, A$ for $\exists x\ (St\ x \wedge A)$

$\forall^{fin} x\, A$ for $\forall x\ (x \text{ finite} \Longrightarrow A)$

$\exists^{fin} x\, A$ for $\exists x\ (x \text{ finite} \wedge A)$

$\forall^{St\ fin} x\, A$ for $\forall^{St} x\ (x \text{ finite} \Longrightarrow A)$

$\exists^{St\ fin} x\, A$ for $\exists^{St} x\ (x \text{ finite} \wedge A)$

where A is any external formula.

The axioms of I.S.T. are :

- all axioms of Z.F.C., restricted to internal formulas. In other words, I.S.T. is an extension of Z.F.C.

- the transfer principle (T) .

Let $A(x, t_1, \dots, t_k)$ be any internal formula without other free variables than $x, t_1, \dots, t_k$ . Then, consider the axiom

$$\forall^{St} t_1, \dots, \forall^{St} t_k\ (\forall^{St} x\, A(x, t_1, \dots, t_k) \Longleftrightarrow \forall x\, A(x, t_1, \dots, t_k))$$

- the idealization principle (I) (with parameters) .

Let $B(U, V, t_1, \dots, t_k)$ be any internal formula with free variables $U, V, t_1, \dots, t_k$ . Then, consider the axiom

$$\forall t_1, \dots, \forall t_k [\forall^{St\ fin} z \exists v (\forall u \in z, B(u, v, t_1, \dots, t_k)$$
$$\Longleftrightarrow \exists v\, \forall^{St} u\, B(u, V, t_1, \dots, t_k)] .$$

- the standardisation principle (S) :

Let $C(z, t_1, \dots, t_k)$ be any external formula, with free variables $z, t_1, \dots, t_k$. Then, consider the axiom

$$\forall t_1, \dots, \forall t_k \, \forall^{St} x \, \exists^{St} y \, \forall^{St} z \, [(z \in y) \Longleftrightarrow ((z \in x) \wedge C(z, t_1, \dots, t_k))] .$$

4) Let us outline a conservativeness proof of IST :

Consider a closed internal formula A and suppose that it has a proof within I.S.T. This proof needs only a finite (in the intuitive sense) number of axioms of Z.F.C., say $A_1, \dots, A_n$, and also some axioms of the schemes (I), (S), (T). We are looking for a proof of A within Z.F.C.

First consider an interpretation $(E, G)$ of $\mathcal{L}_{Z.F.}$ in which the translations $\hat{A}_1, \dots, \hat{A}_n$ of $A_1, \dots, A_n$ are theorems of Z.F.C. (i.e. a model of the theory based on $A_1, \dots, A_n$). In some limit enlargement $({}^*E, {}^*G)$, the interpretations ${}^*\hat{A}_1, \dots, {}^*\hat{A}_n$ are also theorems (by transfer) and, moreover, the interpretations of the schemes (I), (S), (T) are theorems of Z.F.C. if " St x " translates in " $x \in E$ ".

Hence our initial proof translates into a proof of ${}^*\hat{A}$ from ${}^*\hat{A}_1, \dots, {}^*\hat{A}_n$, $({}^*I)$, $({}^*S)$, $({}^*T)$ within Z.F.C.

As A is internal and closed, we have, by transfer, $\hat{A} \Longleftrightarrow {}^*\hat{A}$ (in Z.F.C.) and thus $\hat{A}$ is a theorem in Z.F.C.

But if the model $(E, G)$ is too big, it's not sure that $\hat{A} \Longrightarrow A$ is a theorem of Z.F.C. Fortunately, the finiteness of the family $A_1, \dots, A_n$ allows the existence of a "minimal" model, for which $\hat{A} \Longrightarrow A$ is a theorem (see Nelson's paper for details) ; hence A is a theorem in Z.F.C., which proves the expected conservativeness.

5) Now, there is an algorithm to reduce any external formula $E(t_1, \dots, t_n)$ without other free variables than $t_1, \dots, t_n$ to an internal formula $E'(t_1, \dots, t_n)$ with the same free variables, such that $\forall^{St} t_1, \dots, \forall^{St} t_n \, (E(t_1, \dots, t_n) \Longleftrightarrow E'(t_1, \dots, t_n))$ is a theorem within I.S.T.

In other words, any information that you may formalize within $\mathcal{L}_{I.S.T.}$ by a formula $E(t_1, \dots, t_n)$ is equivalent to the classical property $E'(t_1, \dots, t_n)$, provided the parameters $t_1, \dots, t_n$ are restricted to standard values. This restriction is not important, as we shall see. Notice that the translation $E \longrightarrow E'$ is mechanical ; of course $E'$ may be much more complicated than $E$ and this is why non standard definitions of intuitive concepts may be easier than classical ones.

6) To be standard or not to be. Consider an internal formula $A(x)$ (one free variable only) and assume that $\exists ! x\, A(x)$ is a theorem of Z.F.C. Then, by (T), $\exists^{St} ! x\, A(x)$ is a theorem of I.S.T. ; in other words, any classical object which is defined by a construction in Z.F.C. is standard, when considered as an object in I.S.T. This fact may be compared with its analogue in lesson 2 ("to be natural or not to be").

Clearly non-standard objects of I.S.T. cannot be constructed within Z.F.C. Hence, all that is (in a classical mind) is standard (in a non-standard one). For instance, $\mathbb{N}$, $\mathbb{Q}$, $\mathbb{R}$, $\pi$, the Hilbert space $L^2(\mathbb{R})$, the Grassmann manifold of 5-planes in $\mathbb{R}^{25}$, etc... are all standard.

Now, this may be generalized. Assume that $A(x, t_1, \dots, t_k)$ is an internal formula with free variables $x, t_1, \dots, t_k$ (the latter are "parameters"). If $\forall^{St} t_1, \dots, \forall^{St} t_k \quad \exists ! x\, A(x, t_1, \dots, t_k)$ is a theorem in I.S.T., then by (T), $\forall^{St} t_1, \dots, \forall^{St} t_k \quad \exists^{St} ! x\, A(x, t_1, \dots, t_k)$ is also a theorem in I.S.T.

Thus, any object which is constructed by internal means (i.e. classical constructions) from standard data is standard too.

For instance, for $n$ standard, $\mathbb{R}^n$ is standard ; if $f$ is a standard mapping from $\mathbb{R}^n$ to $\mathbb{R}^m$ ($n, m$ standard), then its image is standard, and standard arguments have a standard image.

This is one reason for which the restriction on standard parameters in § 5 is not important : if you define some concept, it is always about some well-defined mathematical object (e.g. continuity in real analysis), and the latter is standard. The second reason is that whenever you have to prove an internal statement involving parameters $t_1, \dots, t_k$, which sounds like

$\forall t_1, \dots, \forall t_k \, A(t_1, \dots, t_k)$, you may equivalently prove (within I.S.T.)

$\forall^{St} t_1, \dots, \forall^{St} t_k \, A(t_1, \dots, t_k)$ by the transfer principle ; now it is possible to use equivalent forms of this statement within I.S.T.

Thus, as far as I.S.T. is only used as a tool in classical mathematics, we may consider the restriction as immaterial.

7) The standardisation principle has an important consequence, of frequent use in the sequel. We call it

Construction principle : Let $X$ and $Y$ be standard sets and $A(x, y)$ an external formula with free variables $x$ and $y$. Assume that for every standard $x$ $x \in X$, there is a standard $y \in Y$ such that $A(x, y)$. Then there is a standard mapping $f : X \longrightarrow Y$ such that, for every standard $x \in X$, one has $A(x, f(x))$.

This principle yields a lot of standard mappings from external relations ; you have only to satisfy pointwise the criterion $\forall^{St} x \, \exists^{St} y \, A(x, y)$. Notice that the internal properties of $f$ may be proved on the standard values of $x$ only and thus reflect some external properties of $A$.

Proof. If there is a unique standard $y \in Y$ such that ..., use (S) to get a standard subset of $X \times Y$ whose standard elements are those which satisfy $A(x, y)$. This subset is the graph of the expected mapping $f$.

In the general case, use the transfered form of the axiom of choice to reduce the problem to the particular case.

Notice that the construction principle may be written as follows :

$$\forall^{St} x \, \exists^{St} y \, A(x, y) \Longleftrightarrow \exists^{St} f \, \forall^{St} x \, A(x, f(x)) \text{, or dually,}$$

$$\exists^{St} x \, \forall^{St} y \, A(x, y) \Longleftrightarrow \forall^{St} f \, \exists^{St} x \, A(x, f(x)) \text{,}$$

where $f$ is tacitly supposed to be functionnal.

So we have commutation rules for the "external quantifiers" $\forall^{St}$ $\exists^{St}$.

If you remark that the idealization principle (and its dual form) is also a "structural rule" in this "algebra of quantifiers", namely $\exists\ \forall^{St} \Longleftrightarrow \forall^{St\ fin}\ \exists\ \forall$ , you are ready to understand the <u>reduction algorithm</u> mentionned in § 5 .

8) First recall that any formula in $\mathcal{L}_{Z.F.}$ may be put into an equivalent prenex form $Q_1\ x_1\ \dots\ Q_n\ x_n\ B$ where $Q_i$ is $\forall$ or $\exists$ and B is a formula without quantifiers.

A well-known list of the usual transformations is given by Nelson: We reproduce it for a possible use by the reader .

1) $\neg \forall\ x\ A(x) \Longleftrightarrow \exists\ x \neg A(x)$

2) $\forall\ x\ \forall\ y\ A(x\ ,y) \Longleftrightarrow \forall\ y\ \forall\ x\ A(x\ ,y)$

3) $(\forall\ x\ A(x)) \wedge B \Longleftrightarrow \forall\ x\ (A(x) \wedge B)$

4) $(\exists\ x\ A(x)) \wedge B \Longleftrightarrow \exists\ x\ (A(x) \wedge B)$

5) $[(\forall\ x\ A(x)) \Longrightarrow B] \Longleftrightarrow [\ \exists\ x\ (A(x) \Longrightarrow B)]$

6) $[\ A \Longrightarrow (\forall\ x\ B(x))] \Longleftrightarrow [\forall\ x\ (A \Longrightarrow B(x))]$

7) $[(\exists\ x\ A(x)) \Longrightarrow B] \Longleftrightarrow [\forall\ x\ (A(x) \Longrightarrow B)]$

8) $[A \Longrightarrow (\exists\ x\ B(x))] \Longleftrightarrow [\exists\ x\ (A \Longrightarrow B(x))]$

9) $[(\forall\ x\ A(x)) \Longrightarrow (\forall\ y\ B(y))] \Longleftrightarrow [\exists\ x\ \forall\ y\ (A(x) \Longrightarrow B(y))]$

$\Longleftrightarrow [\forall\ y\ \exists\ x\ (A(x) \Longrightarrow B(y)]$

10) $[(\forall\ x\ A(x)) \Longleftrightarrow (\forall\ y\ B(y))]$

$\Longleftrightarrow [(\forall\ x A(x)) \Longrightarrow (\forall\ y B(y)) \wedge ((\forall\ w B(w)) \Longrightarrow (\forall\ z A(z)))]$

$\Longleftrightarrow \exists\ x\ \exists\ w\ \forall\ y\ \forall\ z\ [(A(x) \Longrightarrow B(y)) \wedge (B(w) \Longrightarrow A(z))]$

$\Longleftrightarrow \forall\ y \forall\ z \exists\ x \exists\ w [(A(x) \Longrightarrow B(y)) \wedge (B(w) \Longrightarrow A(z))]$ .

The same rules apply to the external quantifiers $\forall^{St}$ , $\exists^{St}$ ; between internal and external quantifiers, we have the rule

(11) $\forall\ x\ \forall^{St}\ y\ A(x\ ,y) \Longleftrightarrow \forall^{St}\ y\ \forall\ x\ A(x\ ,y)$ and its dual form

$\exists\ x\ \exists^{St}\ y\ A(x\ ,y) \Longleftrightarrow \exists^{St}\ y\ \exists\ x\ A(x\ ,y)$ .

Now consider an external formula $E(t_1\ , \dots\ , t_n)$ with free variables $t_1, \dots, t_n$ (and no others). Using the rules above, we may put all quantifiers on the left, and as the predicate symbol "St" is necessarily associated with a variable, either it is contained in some external quantifier, or it concerns one of the variable $t_1, \dots, t_n$. In any case we get an equivalent formula in prenex form $Q_1x_1 \dots Q_nx_n\, B(x_1, \dots, x_n, t_1, \dots, t_n)$ where $B$ is free of quantifications and internal as regards the variables $x_1, \dots, x_n$ ; here $Q_i$ is one of the quantifiers $\forall$, $\exists$, $\forall^{St}$, $\exists^{St}$.

The reduction algorithm works using the following instructions :

- Push the external quantifiers $\forall^{St}$, $\exists^{St}$ to the left, using rule (11) and the commutation rules deduced from (I) and (S), until no internal quantifier remains on the left of some external one.
- use axiom T or its dual form to get

$$P_1y_1 . P_2y_2 \dots . P_ny_n\, C(y_1, \dots, y_n, t_1, \dots, t_n)$$

with the same free variables as $B$ and only internal quantifiers $P_1, \dots, P_n$, such that

$$\forall^{St} t_1, \dots, \forall^{St} t_n\, E(t_1, \dots, t_n) \iff$$
$$\forall^{St} t_1, \dots, \forall^{St} t_n, P_1y_1 \dots . P_ny_n\, C(y_1, \dots, y_n, t_1, \dots, t_n) .$$

Now forget the eventual "St" before $t_i$, if necessary, and you get the final internal formula $E'(t_1, \dots, t_n)$ such that

$$\forall^{St} t_1, \dots, \forall^{St} t_n\, E(t_1, \dots, t_n) \iff \forall^{St} t_1 \dots, \forall^{St} t_n\, E'(t_1, \dots, t_n) .$$

Of course, this procedure is not the quickest way to get a classical form of some particular external statement. There are some examples in Nelson's paper, but usually you clearly have in mind some intuitive procedure to find the answer and then use the axioms (in a non mechanical way) to prove its validity.

9) Using his algorithm, Nelson proves the following extension of the "standardness law" in § 6.

Let $A(x)$ be an <u>external</u> formula with only $x$ as free variable ; in general, if $\exists ! x\, A(x)$ is a theorem of I.S.T., we cannot conclude that there is a standard $x$ with property $A$. However, <u>given a standard set</u> $V$, <u>we have</u>

$$\exists\, x ! ((x \in V) \wedge A(x)) \Longrightarrow \exists^{St} x((x \in V) \wedge A(x)) .$$

In other words, we cannot select a non standard element in a standard set by means of an external procedure. This means that non standard elements in a standard set have in some sense an "insidious" character : you cannot distinguish one from another !

10) Now, a big shock ! Apply the idealization principle to the formula $(u \in v) \wedge (v \text{ finite})$ ; you get the following theorem : "there is a finite set F such that every standard x is an element of F ".

In our Z.F.F. (Lesson II.3) we had something analogous : the set $\{x \in \mathbb{N}, x \leq \omega\}$ is finite (for $\omega \in \mathbb{N}$ ) and all "naturals" $0, 1, 2, \ldots$ are elements of it. Within I.S.T., these naturals are standard, of course (they "are"). In this light, the reader should not be too much afraid about this finite set F ; indeed, I.S.T. is only a more ambitious answer to the question in II.3, the non formalizable quality "natural" being replaced by the formal predicate "standard".

Although the standard objects are elements of the same finite set F , we cannot infer that there is only a finite number of standard objects ! Indeed, only sets have a "number of elements" and we shall prove that there is no set whose elements are the standard sets and no others.

Such a set S would be finite (every subset of a finite set is finite); but a consequence of axiom (I) and (S) is that a set is standard and finite if and only if all its elements are standard (see also lesson I.9. 10).
Hence the set S would be standard and so $S \in S$ , which contradicts the regularity axiom $S \notin S$ .
More generally, be careful about "the set of all $x \in V$ such that ..." ; if the property is external (and " st x " is the first strictly external formula), this set does not exist in most cases.
Instead we have by (S) "the standard set whose standard elements ...".

11) Let $P(n)$ be an external formula with one free variable n and assume that $P(o)$ and $\forall^{st} n \in \mathbb{N} (P(n) \Rightarrow P(n+1))$ are theorems. Then we have $\forall^{st} n \in \mathbb{N}\ P(n)$ .

This is an external induction principle. Notice that we don't know

whether P is true for non-standard values of $\mathbb{N}$ .
The proof is easy : by (S) , there is a standard subset $E \subset \mathbb{N}$ whose standard elements are those of $\mathbb{N}$ which satisfy P . Now $0 \in E$ and if $n \in E$ , n standard, $n+1 \in E$ . By (T) and usual induction, we infer that $E = \mathbb{N}$ : Thus all standard elements of $\mathbb{N}$ satisfy P .

12) In lesson I.10, we used a permanence principle which is easy to reformulate within I.S.T. with its full strenght, as follows :
Consider a standard set V , an external formula E(x) , an external formula P(x) and an internal formula A(x) , all with x as free variable. Assume the following properties :

i) $\forall x \in V \ (E(x) \Rightarrow P(x))$

ii) $\forall x \in V \ (E(x) \wedge P(x) \Rightarrow A(x))$

iii) $\forall x \in V \ (\neg E(x) \wedge A(x) \Rightarrow P(x))$

iv) There exists no subset of V whose elements are those of V which satisfy E(x) .

Then there exists an $x \in V$ such that $\neg E(x)$ and P(x) are satisfied.

In other words, property P is permanent on some elements which don't satisfy E .

Proof. The set $\{x \in V, A(x)\}$ exists, for A(x) is internal, and contains all $x \in V$ satisfying E(x) . By (iv) it contains elements satisfying $\neg E(x)$ , thus by (iii) we get the conclusion.

Example. $V = \mathbb{N}$ , P(n) is "$u_n \sim 0$" (where u is a given sequence in $\mathbb{R}$ ), A(n) is "$|u_n| < \frac{1}{n}$" and E(n) is "st n" . Thus, if $u_n \sim 0$ for all standard n , there is an infinitely large $n_0$ (recall that any non-standard integer is infinitely large) with $|u_{n_0}| \sim 0$ .
Moreover, take for A(n) "$\forall p \leq n , |u_p| < \frac{1}{p}$" and you get an infinitely large $n_0$ such that $\forall p \leq n_0 , u_p \sim 0$ , which is an important improvement.
Sometimes we have $A(x) = P(x)$ and only conditions (i) , (iv) have to be assumed. For instance, if a sequence is bounded by M for every infinitely large n ,

there is a standard $n_0$ such that for any $n \geq n_0$, $|u_n| < M$.
In both examples, we use that the standard (or non standard) integers satisfy (iv).

13) Exercises.

1) Prove the following statements :

- Two standard sets $X$ and $Y$ are equal iff they have the same standard elements.
- Two standard maps $f, g : X \longrightarrow Y$ (standard sets) are equal iff they take the same values on all standards elements in $X$.
- The numbers $\sqrt{2}$, $e$, $\pi$, the function $e^x$, the field of algebraic real numbers, $\mathbb{R}^{125}$, the projections of a standard point in $\mathbb{R}^{236}$, the sequence $u_n$ such that $u_0 = a$, $u_{n+1} = f(u_n)$ with $a$ and $f$ standard, the sum of a standard series, the dimension of a standard vector space (if any), the empty set, all these objects are standard.
- If $X$ is a standard vector field on a standard manifold $M$ (both smooth), its flow (if any) and the integral curve starting at a standard point is standard ; at standard times it passes at standard points etc...
- If $\omega$ is infinitely large and $n$ standard, $(\mathbb{R}^{\omega})^n$ is not standard.
- If $\omega!$ is infinitely large, so is $\omega$.
- If $E$ is a standard set, so is $\mathcal{P}(E)$, $\mathcal{P}(\mathcal{P}(E))$, etc... and also $\mathcal{P}^n(E)$ for $n$ standard.
- If $f : E \longrightarrow F$ is standard and one-to-one onto, then $E$, $F$ and $f^{-1}$ are standard.
- The union, intersection, product of a standard family of sets is standard (but the family usually contains non standard elements, if not finite).

2) Consider the following statement : "if a real sequence $u_n$ takes infinitesimal values for all infinitely large $n$, there is a standard $n_0$ such that, for every $n > n_0$, $u_n \sim 0$."
- Prove it, using the permanence principle.
- Apply it to the sequence $u_n = \frac{1}{n}$ and conclude that $0 = 1$.
- Find the mistake in your proof.

## 14. A discussion about models of Z.F.C.

In the two last lessons, we introduced models in order to justify I.S.T. in a natural way. Also Nelson points out that, as I.S.T. is consistent with Z.F.C., it has a model which may be used to get "external sets". Now, there is a subtle misunderstanding which may occur about these models : according to the point of view on the foundations of your mathematics, you may either consider that ZFC (and a fortiori IST) has no model, either that it has one !

- In the "idealistic" or "aristotelician" point of view (shared by much logicians), God created the universe (S) of sets - a priori free of contradictions - with all usual properties (including the "choice lemma") which are basic in mathematics ; then men did all the rest, i.e. built algebra, analysis, geometry, and also mathematical logic on (S). For instance languages, theories, models... are described in terms of "sets" ; a particular theory, named $\overline{ZFC}$ (not to be confused with our ZFC), has precisely as axioms the basic properties observed on (S). Two essential results appear, both related with Kurt Gödel's work :

(i) the completeness theorem "if a theory is consistent, it has a model"
(ii) the incompleteness theorem "if a theory extends Peano arithmetics, its consistency cannot be deduced from its axiom system only".

Clearly, models are based on sets of (S) and consistency is an absolute property : a theory is consistent or not, has a model or not.

Now, has $\overline{ZFC}$ a model ? This would imply consistency ; hence, by incompleteness (which applies to $\overline{ZFC}$, of course), a positive answer means that (S) has strictly stronger properties than Zermelo-Fraenkel wanted, thus $\overline{ZFC}$ should be strenghtened and the dog is ruming after its tail... So, no model for $\overline{ZFC}$, as consolation, you learn from ThorwaldSkolem's work that if $\overline{ZFC}$ had a model, it would also have a countable one, a rather paradoxal feature...

- In the formalistic point of view (shared by much mathematicians - e.g. Professor N. BOURBAKI) which is described in this section II, there is no universe (S) behind everything and theories like ZFC are "formal", i.e. "potentially concrete "objects" ; taking ZFC as a basis of mathematics, we no longer ask about its consistency (Following Gödel, it is hopeless to try to prove it with finitary arguments).
Now ZFC (or IST) may be "reflected" as "mathematical theories" $\widehat{ZFC}$ (or $\widehat{IST}$) described in the language $\mathcal{L}_{ZF}$ ; the completeness theorem states that "if a theory is consistent with Z.F. it has a model in ZFC" (note that "$\mathcal{C}$ has a model" is an existential formula of $\mathcal{L}_{ZF}$). Clearly $\widehat{ZFC}$ is consistent with ZF, because axiom (C) can be proved to be

consistent with ZF by finitary arguments, and hence $\widehat{ZFC}$ (also $\widehat{IST}$), has a model in ZFC.

Of course we have no trouble with incompleteness, for everything is relative to the consistency of ZF. Once more : the ingredients of $\overline{ZFC}$ are "sets" of (S), those of ZFC are potential concrete objects (or collections), those of $\widehat{ZFC}$ are described in $\mathcal{L}_{ZF}$ (and you may confuse $\widehat{ZFC}$ and ZFC in the practice, and this leads to different conclusions about models...

The first point of view has an advantage : mathematics is a "natural science" and may discover the laws of a secret world. The risk is that this world may not exist, a possible hard catastrophe ... In the formalistic one, mathematics is only a game, which says nothing about nature. (although it is pleasant to notice a lot of coincidences with experience). Whether the rules of the game are consistent is a problem, but in the bad case it's always time to change the rules - changing an eventual nature is more difficult ...

Fortunately, there is a good reason to avoid civil war within the mathematical community : indeed mathematics is rather an art, with some freedom, neither completely a science, nor completely a sterile game, and whatever foundations you prefer, you only have to be convinced that those of NSA are exactly the same as those of your every day practice of this art.

> "On ne diffère du style d'Archimède que dans les expressions, qui sont plus directes dans notre méthode et plus conformes à l'art d'inventer"
>
> G. W. Leibniz
>
> Théorie des infinitésimaux.

- A classical mathematician : " if $u_n$ tends to $\ell$ , then , for every $\varepsilon > 0$ , there exists an $n_0$ such that for $n > n_0$, ..........., thus $u_n^2$ tends to $\ell^2$.

- A physicist : " if $u_n$ tends to $\ell$ , then , for every infinitely large $n$ , $u_n$ is infinitely close to $\ell$ ; hence $u_n^2$ is infinitely close to $\ell^2$, that is $u_n^2$ tends to $\ell^2$ .

- A non-standard minded mathematician : " By transfer , the problem reduces to the case $u_n$ and $\ell$ standard . Then , for every infinitely large $n$ , $u_n$ is infinitely close to $\ell$ ; hence $u_n^2$ is infinitely close to $\ell^2$, that is $u_n^2$ tends to $\ell^2$ ".

PART III : SOME CLASSICAL TOPICS FROM A NON - STANDARD POINT OF VIEW

## Lesson 1

## GENERAL TOPOLOGY

<u>With enlargements in Z.F.C.</u>

Consider a topological space $(X, \mathcal{O})$, where $\mathcal{O} \subset \mathcal{P}(X)$ is the family of open subsets. In a suitable enlargement, call <u>halo</u> of a point $a \in X$ the subset $h(a) = \bigcap_{\substack{a \in U \\ U \in \mathcal{O}}} {}^*U$ of ${}^*X$.

<u>Theorem</u> (Z.F.C.).

i) $h(a)$ is an $*$-neighbourhood of $a$ ;

ii) all usual topological properties may be expressed by means of this concept.

<u>Within I.S.T.</u>

Consider a standard topological space $(X, \mathcal{O})$ (i.e. $X$ and $\mathcal{O}$ are standard). A point $x \in X$ is called a near standard point $a \in X$ iff every standard open neighbourhood of $a$ contains $x$.

<u>Theorem</u> (I.S.T.).

i) There exists an open neighbourhood of $a$, the points of which are all near $a$ ;

ii) all usual topological properties of standard objects may be expressed by means of the quality "near".

<u>Comments.</u> *0) A suitable enlargement is based at least on $X \cup \mathcal{P}(X) \cup \mathcal{P}(\mathcal{P}(X))$, and some auxiliary sets may be added for particular purposes.

By transfer, we get ${}^*X$, ${}^*\mathcal{O}$ with relations ${}^*\in$, ${}^*\cup$, ${}^*\cap$, ${}^*\subset$ which abusively are written $\in$, $\cup$, $\cap$, $\subset$ (But be carefull about internal subsets - read again Lesson I.9.).

Thus the elements of ${}^*\mathcal{O}$, called $*$-open sets, satisfy the transfered topological properties of $\mathcal{O}$, that is :

- ${}^*X$ is $*$-open,
- ${}^*\emptyset = \emptyset$ is $*$-open,

- every $*$-union of an $*$-family of open sets is $*$-open,
- every $*$-intersection of an $*$-finite family of open sets is $*$-open.

In order to get by transfer the concept of $*$-finite family, we must add to the basis of our enlargement the set $\mathbb{N} \cup \mathcal{P}(\mathbb{N}) \cup \mathcal{P}(\mathcal{P}(\mathbb{N}) \times \mathcal{P}(X))$ . As we have no limitation for this basis, this is not a problem. But we always have to describe it before using enlargements (practically this preliminary step may be forgotten).

0) I.S.T. Here there is no preliminary step, we are in a theory and have only to restrict the play to standard objects. There are no stars, but we recognize similar plays if we compare $X$ to $^*X$ , $\mathfrak{G}$ to $^*\mathfrak{G}$ , standard open sets to transfered open sets $^*U$ .

The topological axioms transfer :

- $X$ is standard open,
- $\phi$ is standard open,
- the union of any standard family of open sets (i.e. a standard subset of $\mathfrak{G}$ ) is standard open,
- the intersection of any finite standard family of open sets is standard open (this family has only standard elements, indeed).

$^*1$) If $\mathfrak{G}$ is infinite, the relation " $U \neq V$ " on $\mathfrak{G}$ is of type $\Gamma_2$ . Hence we get $*$-open sets which are not of type $^*U$ , $U \in \mathfrak{G}$ .

A big part of topology is devoted to properties which are invariant under refinement of open sets. As the intersection of open sets is generally not open, these properties cannot be expressed by means of "minimal" open sets.

Thanks to enlargements, we have the halos, which are intermediate objects leading to direct formulations.

1) IST. Again use idealization to get non standard open sets in $\mathfrak{G}$ . But an essential difference with enlargements is that there is no set of all $x$ near $a$ . The concept of halo fails ! This has a counterpart in ZFC : the halos are not internal sets (recall lesson I.9.) and we know that I.S.T. only describes the play between internal sets .

*2) Let $A$ be a subset of $X$. Define $h(A) = \bigcup_{a \in A} h(a)$ and $H(A) = \bigcap_{\substack{U \supset A \\ U \in \mathfrak{G}}} {}^*U$. Both are $*$-neighbourhoods of $A$, and are usually different. Both could be called halo of $A$. Indeed $H(A)$ is usefull in expressing regularity properties, while $h(A)$ has to do with a big part of elementary topology. Thus we call "halo" the set $h(A)$ and "big halo" the set $H(A)$.

2) IST. Here $x$ is near a standard subset $A$ of $X$, if it is near at least one standard point of $A$.
As an exercise, define a quality corresponding to $H$.

*3) Proof of theorem.
i) The binary relation $(U \in \mathfrak{G}) \wedge (V \in \mathfrak{G}) \wedge (V \subset U) \wedge (a \in U \cap V)$ is of type $\Gamma_2$, due to the axioms of topological spaces. Hence there exists an ideal element $\Omega \in {}^*\mathfrak{G}$ such that $a \in \Omega$ and $\Omega \subset {}^*U$ for every open set $U$ containing $a$. Thus $h(a)$ is an $*$-neighbourhood of $a$.

3) IST.
i) Use idealization to get an open set $\Omega$ - which is not standard in most cases - such that $a \in \Omega$ and every $x \in \Omega$ is near $a$.

*4) and 4) IST. The following table justifies ii) :

| Classical formulation | With enlargements | within I.S.T., for standard objects only |
|---|---|---|
| $A$ finite | ${}^*A = A$ | St $A$, and every element of $A$ is standard. |
| $A$ open | $h(A) \subset {}^*A$ | St $A$, and every point near $A$ is in $A$. |
| $A$ closed | $h(X - A) \subset {}^*X - {}^*A$ | St $A$, and every point near $X - A$ is in $X - A$. |
| $a \in \mathring{A}$ (interior) | $h(a) \subset {}^*A$ | St $A \wedge$ St $a$, and every point near $a$ is in $A$. |
| $a \in \overline{A}$ (closure) | $h(a) \cap {}^*A \neq \emptyset$ | St $A \wedge$ St $a$, and there is a point near $a$ in $A$. |
| $a \in A'$ (limit points) | $h(a) \cap {}^*A \neq \emptyset, \{a\}$ | St $A \wedge$ St $a$, and there is a point near $a$ in $A$ distinct from $a$. |
| $X$ is Hausdorff | $\left.\begin{array}{l} a \neq b \\ a \in X \\ b \in X \end{array}\right\} \Rightarrow h(a) \cap h(b) = \emptyset$ | St $a \wedge$ St $b \wedge (a \neq b)$ implies that no point is near both $a$ and $b$. |

As for properties involving $H$, we have for instance the following : a space $X$ is regular iff $a \in X$, $A \subset X$, $a \notin A$, and $A$ closed implies $h(a) \cap H(A) = \emptyset$ ; a space $X$ is normal iff $A \subset X$, $B \subset X$, $(A \cap B = \emptyset)$, $A$ and $B$ closed imply $H(A) \cap H(B) = \emptyset$. Similar formulation within I.S.T., for $a, X, A, B$ standard. The translation $* \longleftrightarrow$ IST is very easy and in the sequel we leave it to the reader.

The proofs concerning the table above are immediate consequences of theorem (i). We give one example, about " A open".

with $*$ : if $A$ is open and $a \in A$, then (definition of $h$), $h(a) \subset {}^*A$. Hence $h(a) \subset {}^*A$. Conversely, if $h(a) \subset {}^*A$, then ${}^*A$ is an $*$-neighbourhood of $a$ and by transfer $A$ is a neighbourhood of $A$. Hence $A$ is open.

within I.S.T. : Suppose St $A$, St $a$ and $a \in A$. Then, if $A$ is open, it contains every point near $a$. Hence every point near $A$ is in $A$. Conversely, if every point near $a$ is in $A$, the open neighbourhood given by i) is a subset of $A$. Hence $A$ is open.

Notice that two topologies on $X$ with the same halos (or within IST, two standard topologies with the same "near" relation) have the same open sets, hence are the same.

5) To describe continuity with enlargements, we must enlarge the structure based on two spaces $X$ and $Y$ :

Consider $f : X \longrightarrow Y$. Then $f$ is continuous at $a \in X$ iff ${}^*f(h(a)) \subset h(f(a))$.

Proof. Let $V$ be an open subset of $Y$, $f(a) \in V$ ; then $f^{-1}(V)$ is open, hence $h(a) \subset {}^*f^{-1}(V)$ and ${}^*f(h(a)) \subset {}^*V$. Thus ${}^*f(h(a)) \subset h(f(a)) = \bigcap_{\substack{V \text{ open} \\ f(a) \in V}} {}^*V$.

Conversely, if an open $V$ contains $f(a)$, then ${}^*V \supset h(f(a)) \supset {}^*f(h(a))$ ; hence ${}^*(f^{-1}(V)) = ({}^*f)^{-1}({}^*V)$ is an $*$-neighbourhood of $a$ (use theorem i) ) and by transfer $f^{-1}(V)$ is a neighbourhood of $a$.

Within IST, we have the corresponding statement :

A standard mapping $f : X \longrightarrow Y$ between standard topological spaces is continuous

<u>at a standard point</u> $a$ <u>iff</u> $x$ <u>near</u> $a$ <u>implies</u> $f(x)$ <u>near</u> $f(a)$ .

<u>Proof</u>. Exercise.

Now it is clear that an homeomorphism yields a one-to-one correspondance between the halos in both spaces. Same remark for near points.

<u>Remark</u>. In a standard space, we did not define " $x$ near $y$ " for any $y$ , but only for <u>standard</u> $y$ . In metric spaces, we shall see that "near" extends to a relation on the whole space. But in general topology, there is no natural extension, although extensions exist and may be usefull for special purposes.

*6) We may extend to internal objects of an enlarged space $^*X$ the usual topological properties. We get " $*$-properties". For instance, we call an internal $g : {}^*X \longrightarrow {}^*Y$ $*$-continuous if for every $*$-open set $V$ in $^*Y$ , $g^{-1}(V)$ is open in $^*X$ (recall that $V$ is not of type $^*W$ , $W$ open in $X$ , in general). But it is also possible to extend some non-standard formulations to internal (but not standard) objects : we get the corresponding S-notion. To this end, first define the halo of a non standard point $x \in {}^*X$ as the intersection of all $^*U$ , $U$ open in $X$ , with $x \in {}^*U$ . This is not an $*$-neighbourhood (in general) of $x$ ; but we may use this halo to extend definitions. For instance, we say that $g : {}^*X \longrightarrow {}^*Y$ is S-continuous at $x \in {}^*X$ iff $g(h(x)) \subset h(g(x))$ ; but this has nothing to do with $*$-continuity ! (see Lesson I.4.)

6) <u>I.S.T.</u> Here the $*$-properties are the properties themselves, as defined in the classical context. By transfer, they give properties of standard objects. For instance, a standard mapping is continuous iff given any standard open set $V$ , $g^{-1}(V)$ is a standard open set.
To get S-notions, you forget the word "standard" somewhere ; for instance, $g : X \longrightarrow Y$ is S-continuous at $x \in X$ iff every point "near $x$ " (in some extended sense) has its image "near $g(x)$ ". This property is of some importance in metric spaces (see Lesson III.4.).

*7) <u>The shadow</u> $°x$ of a point in $h(X)$ is the set of all $a \in X$ such that $x \in h(a)$ ; the points of $X - h(X)$ have no shadow. If $X$ is a Hausdorff space, $°x$ is a singleton and we call again its unique point $°x$ . In this case, we have a shadow for subsets :

If $A$ is an internal subset of $^*X$ , the shadow $°A$ is the unique subset of $X$ such that every point of $°A$ is the shadow of some point in $A$ ; this $°A$ may be empty, and in any case $^*(°A)$ may be different from $A$ . For instance, in $^*\mathbb{R}$ (with the usual topology), $°(^*\mathbb{R}) = \mathbb{R}$ but what about the shadow of $A = \{r\varepsilon \, , \, r \in \mathbb{Q}\}$ where $\varepsilon$ is some fixed infinitesimal ?

In some cases, we have also a shadow for mappings :

If $Y$ is Hausdorff and $f : {}^*X \longrightarrow {}^*Y$ internal such that $f(X) \subset h(Y)$ , define $°f : X \longrightarrow Y$ by $(°f)(a) = °(f(a))$ .

Further, we shall see that in metric spaces, S-continuity of $f$ has to to do with continuity of $°f$ .

7) IST. We may define the property " a is the shadow of x " as a synonym for " x is near a " ; but "the shadow" is not a set ; however if $X$ is Hausdorff, there is no problem : every near-standard point has a shadow $°x$ . Now, the shadow of a subset $A$ of $X$ is the unique standard subset $°A$ such that every standard element of $°A$ is the shadow of some element of $A$ . Its existence is a consequence of the standardisation principle ; uniqueness follows from the fact that a standard set is characterized by its standard elements.

As for mappings, if $X, Y$ are standard, $Y$ Hausdorff, and if every point of $f(X)$ is near-standard, then $°f$ is the unique standard mapping $°f : X \longrightarrow Y$ such that for every standard $a \in X$ , $f(a)$ is near $(°f)(a)$ . Again use standardisation to get $°f$ .

*8) As far as induced, product or quotient topologies are concerned, the following remarks may be useful :

- if $A \subset X$ and $a \in A$ , the halo of $a$ for the induced topology is $A \cap h(a)$ .
- the halo in the product is the product of the halos.
- the halo in a quotient space is the quotient of the halos in the total space.

8) IST. Replace "halo" by the "near" relation and you get the corresponding statement.

9) Within IST, the characterizations of classical topological properties only concern standard spaces, which is not the case with enlargements. Both approaches are different in nature. But recall that this restriction is immaterial, because

i) All individual spaces which are constructed within Z.F.C. are standard. Thus in $\mathbb{R}$ , $L^2(\mathbb{R})$ , ... , everything works without restriction.
ii) In order to prove a theorem $A(x_1 , \dots , x_n)$ in Z.F.C. ( $x_1 , \dots , x_n$ are the free variables), we may, using the transfer principle, prove in IST the equivalent statement $\forall\, st\, x_1\, \forall\, st\, x_2 \dots \forall\, st\, x_n\, A(x_1 , \dots , x_n)$ ; the latter proof may use any characterization which works only for standard objects.
For instance, let us prove that in a Hausdorff space, every singleton $\{a\}$ is closed. We alternatively prove that in a standard Hausdorff space $X$ , every standard $\{a\}$ is closed. If $x$ is near a standard $b \neq a$ , then $x$ is not near $a$ (use the characterization of "standard Hausdorff") ; thus $x \in X - \{a\}$ , which proves that $X - \{a\}$ is open (char. of "standard open").

10) In this lesson, we compared the " * " and IST non-standard points of view, in order to help the reader in going from one to the other. The " * " approach has as main disadvantage the preliminary need, in any question, of a basic structure to be enlarged ; it would be pleasant to enlarge once for all everything, but this is not possible, due to the non existence of a "set of all sets", within Z.F.C.
On the other hand, within IST this problem disappears ; however, there is a counterpart : we have no external sets, like halos, and we must replace them by some periphrasis.
In his paper, E. NELSON proposes to use a model of IST, for in such a model, external sets make sense ; however, you have to distinguish carefully the membership relation in the model and the membership relation in the theory. This is an occasion of much pitfalls and seems practically not usable.

In the next lesson, we discuss a variant of IST with a concept of external sets. The basic idea is that such external sets as we need are only concerned by very rough operations like unions, intersections, products, etc... that may be clearly described within the language of IST.

11) Exercise. Discuss the following affirmation : Given a set $X$, an enlargement $^*X$, and for each point $a \in X$ a subset $h(a) \subset {}^*X$, there is a topology on $X$ whose halos are these $h(a)$.

---

## Lesson 2

## INTERNAL SET THEORY WITH EXTERNAL SETS

Complete the language $\mathcal{L}_{I.S.T.}$ with a monadic predicate symbol "int" (read " x is internal" for " int x ") and call $\mathcal{L}_{ISTE}$ this new language.

With $\mathcal{L}_{ISTE}$ as underlaying language, consider the theory ISTE based on the following axioms :

1) All axioms of IST restricted to internal sets.

2) The extensionality axiom as in Z.F.

3) The regularity axiom as in Z.F.

4) The subset axiom for any formula and any set.

5) $\forall x \forall y \ ((x \in y) \Longrightarrow \text{int } x)$.

6) $\forall x \ (\text{st } x \Longrightarrow \text{int } x)$.

7) $\exists z \forall y \ ((\text{int } y \Longrightarrow y \in z)$.

Metatheorem. ISTE is a conservative extension of I.S.T. (hence also of Z.F.C.). The translation of any formula $A$ of $\mathcal{L}_{IST}$ into a formula $A'$ of $\mathcal{L}_{ISTE}$ consists in joining "int x" to any variable $x$ occuring in $A$, and then $A'$ is a theorem within ISTE if and only if $A$ is a theorem within I.S.T.

<u>Comments</u>. 1) In the extension ZFC $\longrightarrow$ IST, all axioms of Z.F.C. remained valid, when applied to formulas not including the predicate symbol " st " ; in that case, the translation $A \longrightarrow A'$ from $\mathcal{L}_{ZF}$ into $\mathcal{L}_{IST}$ was simply $A \longrightarrow A'$ .
Here there is a little difference : the axioms of IST remain valid provided any variable is quoted with "int".

2) Call <u>external</u> the sets of ISTE. Thus $\neg$int x means that x is strictly external.
Axioms (2) to (7) apply to all external sets (notice that (2) , (3) , (4) apply to internal sets due to (1).).
As a first consequence, we get an unique set $\mathbb{I}$ , whose elements are all internal sets (apply (4) to the property "int y" working on the set z of axiom (7) and use (2) for uniqueness.).
Also, we get a subset $\mathbb{S}$ of $\mathbb{I}$ whose elements are all standard sets (property " st y ".).
From axiom (5) we see that the membership relation $\in$ only has to work between internal and external sets ; an external set has no external elements !
Notice some analogy with the theory of classes (Von Neumann-Bernays).

3) Among others, the following operations are defined within ISTE, due to axiom (4) :

$$\complement_y x = \{z \in y \; , \; z \notin x\} \; ;$$

$$x \cup y = \{z \in \mathbb{I} \; , \; (z \in x) \vee (z \in y)\} \; ,$$

$$x \cap y = \{z \in \mathbb{I} \; , \; (z \in x) \wedge (z \in y)\} \; ,$$

$$\bigcup_{y \in x} y = \{z \in \mathbb{I} \; , \; \exists\, y \in x \; , \; z \in x\} \; ,$$

$$\bigcap_{y \in x} y = \{z \in \mathbb{I} \; , \; \forall\, y \in x \; , \; z \in y\} \; ,$$

$$x \times y = \{z \in \mathbb{I} \; , \; (\exists\, u \in x) \wedge (\exists\, v \in y) \; , \; z = (u \, , v)\} \; .$$

(Recall that $(u , v) = \{\{u\} , \{u , v\}\}$ which makes sense for internal sets u and v.)

As usual $x \subset y$ means $\forall z ((z \in x) \Longrightarrow (z \in y))$ . But the following consequence is

unusual : for every $x$ , we have $x \subset \mathbb{I}$ (axiom (5)), but also, for any internal $x$ , $x \in \mathbb{I}$ ; moreover we have $\mathbb{I} \times \mathbb{I} \subset \mathbb{I}$ ...
A binary relation is a subset of a product $x \times y$ ; a mapping $f : x \longrightarrow y$ is a relation satisfying the usual functionality properties. Such a relation is internal if the corresponding subset of $x * y$ is internal. Clearly, the image of an internal subset of $x$ under an internal mapping is internal (use axiom (1)). More generally, <u>all Z.F. set theoretic operations on internal sets have internal results</u>.

4) From axiom (4), we may obtain strictly external sets, whenever in I.S.T. we get nothing. For instance, $\mathbb{S}$ <u>and</u> $\mathbb{I}$ <u>are strictly external</u> (if not, we have $\mathbb{I} \in \mathbb{I}$ , which contradicts axiom (3) ; as for $\mathbb{S}$ , if $\mathbb{S}$ is internal, so is $\mathbb{S} \cap \mathbb{N}$ for $\mathbb{N}$ is standard, and we know that the standard elements of $\mathbb{N}$ have no last element, which would be the case for an internal subset of $\mathbb{N}$ .)
Similarly the sets $x_s = x \cap \mathbb{S}$ are in most cases strictly external.

5) Any concept which needs " $\forall x \exists y (x \in y)$ " is excluded in ISTE, whenever strictly external sets are involved. Indeed, this would imply $\forall x (x \in \mathbb{I})$ by axiom (5), hence $\mathbb{I} \in \mathbb{I}$ which contradicts axiom (3).

Thus, if $a , b , c , \ldots$ are not internal, there is no set $\{a , b , c, \ldots\}$ , particularly no singleton $\{a\}$ . Of course, the power set $\mathcal{P}(x)$ is only defined for $x$ internal.

6) What about finite sets within ISTE ?
Again a set is transitive if $\forall y \in x ((z \in y) \Longrightarrow (z \in x))$ ; an ordinal is a transitive set whose elements are all transitive. Hence all ordinals we had in IST appear also in ISTE as internal ordinals. However, there are also external ordinals !
For instance, $\mathbb{N}_s$ is transitive, because if $y \in \mathbb{N}_s$ and if $z \in y$ , then $z \in \mathbb{N}_s$ (a standard finite set has all its elements standard) ; now every element of $\mathbb{N}_s$ is a standard ordinal, thus transitive. So $\mathbb{N}_s$ is a strictly external ordinal.
At this stage, a vehement discussion begins between $\mathbb{N}$ and his subset $\mathbb{N}_s$ ; each claims that he is the highest ! Unfortunately, if $\mathbb{N}_s \in \mathbb{N}$ , we get $\mathbb{N}_s$ internal ;

if $\mathbb{N} \in \mathbb{N}_s$ , we get $\mathbb{N} \in \mathbb{N}$ and if $\mathbb{N}_s = \mathbb{N}$ , there is no non-standard analysis. All this being excluded, everybody is satisfied ! Indeed, <u>external ordinals cannot be totally ordered.</u>

On the other hand, as singletons are excluded, an external set has no successor ; thus <u>the concept of finite or limit ordinal has no sense for external sets</u>.

7) Now we understand better why the statement "there exists a finite internal set $F$ such that every standard set is an element of $F$ " (see lesson II. 7.§ 10) has not the unexpected consequence that standard objects are in finite number : we have $\mathfrak{S} \subset F$ but as $\mathfrak{S}$ is external, we cannot compare it with some external "finite" ordinal, for we don't know what this means !

8) As an illustration of ISTE, let us consider a standard topological space $(X, \mathfrak{G})$ .

We have the (external) set $X_s$ and if $a \in X_s$ , the halo $h(a)$ is the external set $\bigcap\limits_{\substack{U \in \mathfrak{G}_s \\ a \in U}} U$ .

The union of all $h(a)$ , $a \in X_s$ is not defined, for the family of all $h(a)$ is not a set (its elements would be external) ; but we may define

$$h(X) = \{x \in X \, , \, \exists \, a \in X_s \text{ with } x \in h(a)\}$$

and call it the halo of $X$ .

Thus $X$ is Hausdorff iff " $\forall\, a \in X_s$ , $\forall\, b \in X_s$ , $(a \neq b) \Longrightarrow h(a) \cap h(b) = \emptyset$ ", etc... (re-write the characterizations of Lesson 1).

As you remark, we have nearly the same concepts as with enlargements, but also the advantages of IST.

Of course, some care is necessary ! For instance, if you consider the integral of some function on the halo of a point, you may find astonishing results, because this operation cannot be extended to external domains with its usual properties.

9) <u>Proof of the metatheorem</u>. Consider a model of IST within ZFC based on a set $M$ ; the graph of $\in$ is a subset $G \subset M \times M$ and the graph of "st" is a subset $\Sigma \subset M$ .

Let $\varphi : M \longrightarrow \mathcal{P}(M)$ be defined by $\varphi(x) = \{z \in M \, , \, (z \, , x) \in G\}$ : Then $z \in \varphi(x) \Longleftrightarrow (z \, , x) \in G$ (theorem of ZFC) and after the extensionality axiom, $\varphi$ is one-to-one. We use this to get an interpretation of $\mathcal{L}_{ISTE}$ on the basic set $P = \mathcal{P}(M)$ as follows :

the graph of $\in$ is $\{(\varphi(z) \, , X)$ such that $z \in X\} \subset P \times P$ ,

the graph of $=$ is the diagonal of $P \times P$ ,

the graph of st is $\{\varphi(x) \, , x \in \Sigma\} \subset P$ ,

the graph of int is $\{\varphi(x) \, , x \in M\}$ .

It is easy to check that this interpretation is a model of ISTE. Hence ISTE is consistent.

To prove that, indeed, it is a conservative extension of IST, we may outline a finitary proof (but we could also use the fact that the model above is in some sense "minimal").

Let A be a closed formula in $\mathcal{L}_{IST}$ , A' its translation in $\mathcal{L}_{ISTE}$ . Assume that A' is a theorem of ISTE. Then, any external set occuring in the proof of A' may be described as a subset of some internal set defined by the translation of some external formula of $\mathcal{L}_{IST}$ (the only "providing" axioms are (1) and (4) ; other axioms are just enough restrictive to make this remark true) ; from this we get a proof of A within IST , which proves conservativeness.

Lesson 3

COMPACTNESS

THEOREM (I.S.T.E.). A standard subset A of a standard Hausdorff space $(X, \mathfrak{G})$ is compact iff $A \cap h(A) \supset A$.

Comments. 1) The theorem asserts that every point $x$ in $A$ has a (unique) shadow $°x$ in the external set $A_s$ : we get a kind of external fibration of $A$ over $A_s$ which fibers are the halos of the standard points of $A$, intersected with $A$.

2) Before proving the theorem, which was one of the first important characterization discovered by A. Robinson, let us give some immediate applications.

- Every closed interval $[a, b]$ in $\mathbb{R}$ is compact. By transfer, we may assume $a, b$ standard ($\mathbb{R}$ and its usual topology are standard, of course). Now if $a \leq x \leq b$, we know (see lesson I.11) that $x$ has a shadow such that $a \leq °x \leq b$. Hence $[a, b]$ is compact.

- Every compact subset $A$ in a Hausdorff space $X$ is closed. By transfer, we may assume $A$ and $X$ standard. If $b \in (X-A)_s$ and $a \in A_s$, we have $h(a) \cap h(b) = \emptyset$ ; hence $h(b) \cap h(A) = \emptyset$ and $h(b) \subset X-A$, for $A \subset A \cap h(A)$ ; thus $X-A$ is open (see table in lesson III.1).

- The product of two compact spaces $X$ and $Y$ is compact. By transfer, we may assume $X$ and $Y$ standard. But in this case, we know that $h(X \times Y) = h(X) \times h(Y)$. Thus $h(X \times Y) = X \times Y$. The converse is also clear.

- BOLZANO-WEIERSTRASS Lemma. Every infinite subset $B$ in a compact space $X$ has a limit point (i.e. an accumulation point).

By transfer, we may assume $X$ and $B$ standard. Then (see III.1) there is an $\alpha$ in $B-B_s$ ; it has a shadow $a$, by compactness of $X$. Hence $\alpha \in h(a) \cap B$ and $\alpha \neq a$ (for $\alpha$ is not standard). Thus (characterization of limit points) a is a limit point of B.

Remark. This procedure yields all limit points of B.

3) Exercises. Prove that

- a closed subset in a compact space is compact ;
- a compact space has a compact image under a continuous mapping ;
- a continuous one to one mapping from a compact space onto a Hausdorff space has a continuous inverse.

4) Proof of the theorem. We can apply the reduction algorithm in Lesson II to the statement " $A \cap h(A) \supset A$ " and get the usual definition of compactness. But the reader could object that he is not clever enough to perform this formal trick, and that he wants a more "human" proof, where the ideas appear. Here is one : Assume $A$ standard compact in $X$ ; if there is an $\alpha \in A - h(A) \cap A$, we get for every $a \in A_s$ a standard open neighbourhood $V(a)$ of $a$ such that $\alpha \notin V(a)$. The construction principle (lesson II.7, § 8) yields a standard map $\tilde{V} : A \longrightarrow \mathfrak{G}$ (recall that $\mathfrak{G}$ is standard) such that, for every $a \in A_s$, $a \in \tilde{V}(a)$ and $\alpha \notin \tilde{V}(a)$ ; this is a standard open covering of $A$ (use transfer principle to prove that every $x \in A$ is covered) and by compactness, there is a finite standard subset $F \subset A$ such that $B = \bigcup_{a \in F} \tilde{V}(a) \supset A$. But (note that every element in $F$ is standard) $\alpha \notin B$, for $\alpha \notin \tilde{V}(a)$ if $a \in F$. This is a contradiction ; thus $A \subset A \cap h(A)$.

Conversely, assume $A$ standard non compact. Let $\mathfrak{R}$ be a standard open covering of $A$ without finite standard subcover. The idealization principle yields an $\alpha \in A$ which is contained in no standard element of $\mathfrak{R}$ ; hence $\alpha \in h(A)$ and $A \not\subset A \cap h(A)$.

5) The present characterization of compactness is certainly an important tool in topology, if we want to use Nonstandard Analysis. But sometimes we use directly the fact that a set has all its points near-standard (that is near some standard point) without referring to compactness ; for instance to prove uniform continuity of a continuous (standard) function $f$ on a (standard) interval

$[a, b]$ as in lesson I.3, we use $\overset{\circ}{x}$ for $x \in [a, b]$, which needs not a compactness argument.

6) Exercise. Call $X$ the intervall $[0, 1]$ in $\mathbb{R}$ and $\mathcal{B}$ the family of cofinite subsets of $X$ (i.e. $A \in \mathcal{B}$ iff $X - A$ is finite). This is a basis for a topology $\mathcal{O}$ on $X$.

i) Prove that $(X, \mathcal{O})$ is compact and standard ;

ii) Prove that the identity $f(x) = x$ of $[0, 1]$ with the real topology onto $X$ is continuous, but yet not a homeomorphism ;

iii) Compare the halos for both topologies and conclude that they are not homeomorphic.

---

## Lesson 4

## METRIC SPACES WITHIN ISTE

Let $(X, d)$ be a standard metric space. The following external sets may be used to describe metric properties :

- the sets $X_s$ of all standard points,
- the set $X_a$ of all approachable points, i.e. $\{x \in X, \exists y \in X_s, d(x, y) \sim 0\}$,
- the set $X_f$ of all finite points, i.e. $\{x \in X, \forall y \in X_s, d(x, y) \text{ finite}\}$,

clearly $X_s \subset X_a \subset X_f$ ;

- the set $X - X_a$ of unapproachable points and its subset $\{x \in X, \exists \text{ st } \varepsilon > 0, \forall y \in X_s, d(y, x) > \varepsilon\}$ of uniformly unapproachable points.

PROPOSITION 1. A standard subset $A$ of $X$ is bounded if and only if $A \subset X_f$.

PROPOSITION 2. $(X, d)$ is complete if and only if every unapproachable point is uniformly unapproachable.

PROPOSITION 3. The topology associated with d is characterized by the property $h(a) = \{x \in X , \ d(x , a) \sim 0\}$ for every standard $a \in X$ .

Comments. 0) By a standard metric space we mean a standard set $X$ together with a standard mapping $d : X \times X \longrightarrow \mathbb{R}$ . Then, for $x$ and $y$ standard, $d(x , y)$ is standard. In the sequel we write $x \sim y$ instead of $d(x , y) \sim 0$ . Thus $\sim$ is an external equivalence relation and proposition 3 means that " $x$ near-standard" and " $x \sim a$ with $a$ standard" is the same property.

The five external sets above don't make sense within I.S.T., but their defining external properties do within its language ; the use of external sets leads to a shorter exposition but is not essential in the proofs ; you may avoid it and replace in the sequal all halos, $X_s$ , $X_a$ , $X_f$ , etc by the properties of their individuals.

1) In some cases, we have $X_a = X_f$ ; an example is $\mathbb{R}$ with the usual distance $|x - y|$ , for every finite real has a shadow. Note that here all unapproachable points are at infinitely large distance from standard points, hence they are uniformly unapproachable.

However if we consider the distance $\dfrac{|x - y|}{1 + |x - y|}$ on $\mathbb{R}$ , the relation $\sim$ , $X_a$ and the topology are the same as before, but $X_f = X$ .

2) We may rewrite in the present context the non-standard characterization of limits, Cauchy sequences, etc... For instance, a standard sequence $u_n$ in the standard metric space $(X , d)$ is a Cauchy sequence iff, for every infinitely large integer $p$ and $q$ , $u_p \sim u_q$ .

3) Proof of Proposition 1. Suppose $A \subset X$ standard and bounded ; then there is a standard bound $\rho \in \mathbb{R}$ such that $A$ is contained in the standard open ball $B(a , \rho)$ where $a \in X_s$ ; hence every point in $A$ is at finite distance from $a$ , and also from all other standard finite points. Thus $A \subset X_f$ .

Conversely, if $A$ is standard and $A \subset X_f$ , consider an open ball $B(a , \rho)$ with $a \in X_s$ and $\rho$ infinitely large. We have $A \subset X_f \subset B(a , \rho)$ . Thus

A is bounded (note that by transfer, there exists a standard $\rho$ with $A \subset B(a, \rho)$ ).

4) Proof of Proposition 2. Suppose $(X, d)$ complete and $x$ unapproachable. If $x$ is not uniformly unapproachable, there exists for every standard $\varepsilon > 0$, $\varepsilon \in \mathbb{R}$, a point $a \in X_s$ such that $d(a, x) < \varepsilon$. Applying the construction principle (see lesson II.7), we get a standard sequence $a_n$ such that for every $n \in \mathbb{N}_s$, $d(a_n, x) < \frac{1}{n}$ ; this is a Cauchy sequence and its limit is a standard point $b$ such that $d(b, x) \sim 0$. Thus $x$ is approachable, which is a contradiction.

Conversely, consider a standard Cauchy sequence $a_n$ and an infinitely large $\omega \in \mathbb{N}$. Suppose that $a_\omega$ is unapproachable ; then, for every standard $\varepsilon > 0$, there is a $p \in \mathbb{N}_s$ such that $d(a_p, a_\omega) < \varepsilon$ (Cauchy's condition) ; but $a_p$ is standard, and therefore $a_\omega$ is not uniformly unapproachable. This contradicts the hypothesis ; hence $a_\omega \sim x$ for some standard $x$, which thus is a limit of the sequence.

By transfer, every Cauchy sequence has a limit.

5) Proof of Proposition 3. For any $a \in X_s$, the halo $h(a)$ is the (external) intersection of all standard open neighbourhoods of $a$, hence also the intersection of all standard open balls centered at $a$ ; on the other hand $x \sim a$ means that $x$ is contained in all these balls. This proves proposition 3.

6) Recall that every theorem in Z.F.C. about metric spaces is equivalent to the corresponding statement restricted to standard data ; to get a proof, we may then use the properties above as intermedia. We give some examples.

- a metric space is Hausdorff. First transfer : a standard... To prove it, take $a, b \in X_s$ and $a \neq b$. Then $d(a, b)$ is standard $> 0$, therefore not infinitesimal. By proposition 3, we get $h(a) \cap h(b) = \emptyset$ which is characteristic for standard Hausdorff spaces (see lesson 1).

- a compact metric space is complete. First transfer : a standard... As $h(X) = X$, every point is approachable. Conclude by means of proposition 2.

Exercises. Prove that a compact subset in a metric space is bounded.

• Prove that on a standard set $X$, two standard distances $d$ and $d'$ are equivalent in the usual sense iff for every $x, y \in X$, $x \neq y$, both numbers $\frac{d(x,y)}{d'(x,y)}$ and $\frac{d'(x,y)}{d(x,y)}$ are finite.

7) A metric space where every infinite subset has a limit point is compact (converse of Bolzano-Weierstrass lemma).

The usual proof of this important result is quite a mouthful. Here we follow A. Robinson's nice proof in [R], rewriting it within ISTE.

First transfer : A standard... every standard...

Call the space $X$ and consider an eventual $\alpha \in X - h(X)$. Then,

- either $\alpha$ is uniformly unapproachable, i.e. " $\exists$ st $\varepsilon > 0$,

$\forall$ st $y$ $(y \in X \Rightarrow d(\alpha, y) > \varepsilon)$ "

which following axiom (I) is equivalent to

" $\exists$ St $\varepsilon > 0$, $\forall$ St fin $F$, $F \subset X$, $\exists\, y \in X_s$ $(\forall\, x \in F,\ d(x,y) > \varepsilon)$ ".

From the construction principle, we get a standard map $f : \mathcal{P}_y(X) \longrightarrow X$ ($\mathcal{P}_y(X)$ is the standard set of internal subsets of $X$ containing $y$) such that, for every standard finite $F \subset X$ and every $x \in F$, $d(x, f(F)) > \varepsilon$. By induction, we get a standard sequence $u_n$ such that $u_0 \in X_s$ and $u_{i+1} = f(\{u_j, j \leq i\})$. Thus $d(u_i, u_j) > \varepsilon$ for $i \neq j$ (use transfer to prove it for all $i, j$) and the set of its values is infinite without limit point, which is a contradiction.

- either $\alpha$ is not uniformly unapproachable, i.e. for every $n \in \mathbb{N}_s$, there is an $x \in X_s$ with $d(\alpha, x) < \frac{1}{n}$. The construction principle yields a standard sequence such that $d(\alpha, u_n) < \frac{1}{n}$ for every standard $n$ ; it has a standard limit point $a$ and for some $p \in \mathbb{N}_s$, we have $d(\alpha, a) > \frac{1}{p}$ (for $\alpha \notin h(a)$) ; but for some $q > 2p$ holds $d(u_q, a) < \frac{1}{2p}$ and we get the contradiction $\frac{1}{p} < d(a, \alpha) \leq d(a, u_q) + d(u_q, \alpha) < \frac{1}{2p} + \frac{1}{q} < \frac{1}{p}$. Thus $X = h(X)$, which proves compactness of $X$.

8) How to be continuous. Consider an internal mapping $f : X \longrightarrow X'$ between two standard metric spaces $(X, d)$ and $(X', d')$. As usual, $f$ is continuous at a point $x \in X$ if the famous condition " $\forall \varepsilon \in \mathbb{R}^+, \exists \eta \in \mathbb{R}^+,$ $d(x, y) < \eta \Longrightarrow d'(f(x), f(y)) < \varepsilon$" is satisfied. But our intuition suggests condition " $\forall y \in X, y \sim x \Longrightarrow f(y) \sim f(x)$ " which we call S-continuity at $x$. Both conditions are independant - recall $\sin \omega xy$ in section I - but from lesson III.1 and proposition 3 above, we infer that

a standard mapping $f$ is continuous at a standard point $x$ if and only if it is S-continuous at $x$.

Recall that a standard $f$ is continuous on $X$ iff it is continuous at every standard point of $X$ (transfer). But what's going on whenever it is S-continuous at every point of $X$ ? Rewrite lesson I.3 in the general context of standard metric spaces and you get uniform continuity.

For a non-standard $f$ these characterizations fail, but we get a substitute in case $f(X_s) \subset h(X')$ (that is, the image of every standard point is near standard in $X'$; it has an unique shadow) : the construction principle yields an unique standard mapping $^\circ f$ - called the shadow of $f$ - such that for every $x \in X_s$, $(^\circ f)(x) = {}^\circ(f(x))$. (Note that in case $f$ is standard, $^\circ f = f$ by uniqueness) ; now we have the following proposition .

PROPOSITION 4. i) If $f$ is S-continuous at every point of $X_s$ then $^\circ f$ is continuous on $X$ ;

ii) If $f$ is S-continuous at every point of $X$ then $^\circ f$ is uniformly continuous on $X$ .

We leave the proof to the reader, as an exercise. This proposition in an important tool in the sequel ; it yields standard continuous mappings as shadows of non standard ones. The corresponding classical technic is to use convergent sequences, which often is somewhat cumbersome. In the next lesson, we see why shadows replace efficiently sequences.

Lesson 5

# FUNCTIONNAL SEQUENCES

Let $(f_n)$ be a standard sequence of mappings of a standard metric space $(X, d)$ into a standard metric space $(X', d')$.

PROPOSITION 1. i) $f_n$ <u>converges pointwise to a standard mapping</u> $f$ <u>if and only if, for every</u> $x \in X_s$ <u>and every infinitely large</u> $n \in \mathbb{N}$, $f_n(x) \sim f(x)$.

ii) $f_n$ <u>converges uniformly to</u> $f$ <u>if and only if, for every</u> $x \in X$ ...

PROPOSITION 2. <u>If</u> $X$ <u>and</u> $X'$ <u>are compact and if for some infinitely large</u> $\omega \in \mathbb{N}$, $f_\omega$ <u>is</u> S-<u>continuous at every standard point of</u> $X$, <u>the sequence</u> $f_n$ <u>has a standard subsequence which converges uniformly to an uniformly continuous limit</u> $f$.

PROPOSITION 3. <u>If</u> $X$ <u>and</u> $X'$ <u>are compact and if the sequence</u> $f_n$ <u>is uniformly equicontinuous, proposition</u> 2 <u>works for every infinitely large</u> $\omega$.

<u>Comments</u>. 0) The limit $f$ of a standard sequence is clearly standard, for it is unique. Note the difference between both types of convergence : in one case we have "every standard $x$ " while uniformity needs "every $x$ ".

The proof of proposition 1 is an easy exercise, left to the reader.

In case of pointwise convergence, we have $f = {}^\circ(f_\omega)$ for any infinitely large $\omega$. This has no incidence on $d'(f(x), f_\omega(x))$ for non standard $x$. Now suppose the mappings $f_n$ continuous. We claim that $f$ is continuous. Indeed consider $a \in X_s$. For every standard $n$, we have $f_n(a) \sim f_n(x)$. By permanence principle (see II.7), there is an infinitely large $\omega$ with $f_\omega(a) \sim f_\omega(x)$ (use the "trick" $d'(f_n(x), f_n(a)) < \frac{1}{n}$ for $n \in \mathbb{N}_s$ ). Thus $f(x) \sim f_\omega(x) \sim f_\omega(a) \sim f(a)$, which ends the proofs. By transfer, we put standard away before $X, Y, f_n, f$ and get the classical result about continuity versus uniform limits.

1) To prove proposition 2, we use proposition 4 in lesson 4. As $X'$ is compact, the mapping $f_\omega$ has a shadow $f = {}^\circ(f_\omega)$ . We built an uniformly convergent subsequence as follows : $f$ is continuous on $X$ , hence uniformly continuous (compactness of $X'$ ) ; hence for every standard $\varepsilon > 0$ and $\nu \in \mathbb{N}_s$ , the following statement is true " $\exists\, n > \nu$ , $\forall\, x \in X$ , $d'(f(x), f_n(x)) < \varepsilon$ " (take $n = \omega$ ). By transfer, there is a standard $n$ with this property ; thus induction and construction principle yield a standard sequence $n_p$ such that, for every $p \in \mathbb{N}_s$ , we have $\forall\, x \in X$ , $d'(f(x), f_{n_p}(x)) < \frac{1}{p}$ ; by transfer, this is true for every $p \in \mathbb{N}$ , which proves uniform convergence of $f_{n_p}$ to $f$ .

2) Equicontinuous means formally

"$\forall\, \varepsilon > 0$ , $\exists\, \eta > 0$ , $\forall\, x \in X$ , $\forall\, y \in X$ , $\forall\, n \in \mathbb{N}$ , $d(x,y) < \eta \Longrightarrow d'(f_n(x), f_n(y)) < \varepsilon$".

In our case the sequence is standard and by transfer, we get

"$x \sim y \quad > f_n(x) \sim f_n(y)$" for any $n$ , which proves S-continuity of any $f_\omega$ .

This is proposition 3.

3) Proposition 2 and 3 imply Ascoli's lemma, after transfer: But where usually you need Ascoli to get some mapping $f$ , that is a solution of a given problem, you may instead directly consider the shadow of an $f_\omega$ (after transfer of the problem) and prove its accuracy.

Indeed, we can conclude even in case Ascoli doesn't apply: For instance consider a standard sequence of piecewise constant functions on $[0,1]$, taking the values $\lambda_n^i$ on $[a_n^i, a_n^{i+1}[$ , where $\lambda_n^i$ and $a_n^i$ are standard sequences: Assume that $\lim_{n\to\infty} (\sup_i |a_n^{i+1} - a_n^i|) = 0$ and that there is a constant $m$ with $|\lambda_n^{i+1} - \lambda_n^i| < M\, |a_n^{i+1} - a_n^i|$ for every $n, i$ . Then for every infinitely large $\omega$ , $f_\omega$ has a continuous shadow and proposition 2 applies.

# Lesson 6

## SOME EXERCISES TO GET SUPPLE

We use here on some examples the material of lesson 1 to 5 in this section: The reader who is able to reproduce the proofs we give is ready for more substancial topics...

1. All norms on $\mathbb{R}^n$ are equivalent.

• Transfer : "all standard norms... " (don't forget $n$ standard) ;

• Let $\{e_i\}$ be a standard basis of $\mathbb{R}^n$ , $\|\Sigma x_i e_i\| = \sup_i |x_i|$ the standard norm we use as reference and $N$ a standard norm. We have to prove that, for every $x \in \mathbb{R}^n$ , $x \neq 0$ , $\frac{N(x)}{\|x\|}$ and $\frac{\|x\|}{N(x)}$ are finite.

On one hand, $\frac{N(x)}{\|x\|} \leq \Sigma N(e_i)$ which is standard, hence finite. On the other hand, let $S$ be the unit sphere fore $\| \|$ , and $x = \Sigma x_i e_i$ a point on $S$ ; we have $|x_i| \leq 1$ for every $i$ , and at least one of them, say $|x_1|$ , equals 1. Then $x$ has a shadow $t = \Sigma t_i e_i$ with $^\circ x_i = t_i$ and $t_1 = \pm 1$ . Hence $N(t)$ is standard $> 0$ and $N(t) \leq N(x) + \|t - x\| \Sigma N(e_i) \sim N(x)$ ; thus $N(x)$ is not infinitesimal, which ends the proof.

2. The Hilbert cube is compact.

The Hilbert cube $C = \{u : \mathbb{N} \to \mathbb{R} , \forall i \geq 1 , |u_i| \leq \frac{1}{i}\}$ is clearly standard (cf. II.7 "to be standard or not to be") ; hence compactness for the norm $\sqrt{\sum_1^\infty u_i^2}$ means that every $u \in C$ has a shadow in $C$ .

Of course, for every standard index $i$ , $u_i$ has a shadow in $\mathbb{R}$ with $|^\circ u_i| \leq \frac{1}{i}$ . The construction principle yields from this fact a standard sequence $v : \mathbb{N} \to \mathbb{R}$ such that for every standard $i$ , $v_i \sim u_i$ and thus $|v_i| \leq \frac{1}{i}$ (for $|u_i| \leq \frac{1}{i}$ and $v_i$ is standard) ; by transfer, we have $|v_i| \leq \frac{1}{i}$ for every $i$ $i \in \mathbb{N}$ , which implies $v \in C$ .

We prove that $\sum_1^\infty (u_i - v_i)^2 \sim 0$ , i.e. that $v = {}^\circ u$ in the normed space $L_2$

Consider a standard $\varepsilon > 0$ ; then, for every standard $n$ , we have $\sum_1^n (u_i - v_i)^2 < \frac{\varepsilon}{2}$ for $u_i \sim v_i$ , $i \leq n$ . This internal property is permanent until some infinitely large $\omega$ and we get

$$\sum_1^\infty (u_i - v_i)^2 < \frac{\varepsilon}{2} + \sum_{\omega+1}^\infty (u_i - v_i)^2 < \frac{\varepsilon}{2} + 4 \sum_{\omega+1}^\infty \frac{1}{i^2} < \varepsilon ,$$

because $\sum_{\omega+1}^\infty \frac{1}{i^2} \sim 0$ (recall that $\sum_1^\infty \frac{1}{i^2}$ is convergent !).

3. From Sperner's lemma to Brouwer's fixed point theorem.

Let $\Delta$ be a (standard) triangle in $\mathbb{R}^2$ and $f : \Delta \longrightarrow \Delta$ a (standard) continuous mapping. Then $f$ has a (standard) fixed point.

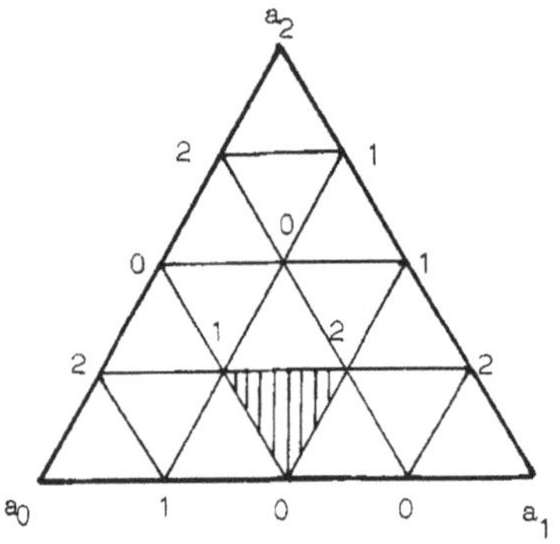


The picture clearly suggest what is a subdivision of $\Delta$ in $4^n$ similar subtriangles, for every $n \in \mathbb{N}$ .

For a fixed $n$ , let $\psi_n$ be any (internal) mapping from the set of vertices of the n-th subdivision into the set $\{0 , 1 , 2\}$ , such that every vertex on the edge $a_i a_j$ takes value $i$ or $j$ .

Sperner's lemma, a purely combinatorial result, insures that there is at least one triangle in the subdivision of type $(0 , 1 , 2)$ .

Now, put $A_i = \{x \in \Delta , \; d(f(x) , a_i) \geq d(x , a_i)\}$ . By induction and construction principle, there is a standard sequence $\psi_n$ of mappings of the type above such that $\varphi(x) = i$ implies $x \in A_i$ (use the fact that a vertex on the edge $a_i a_j$ cannot come closer to both $a_i$ and $a_j$ under $f$ ). For $n$ infinitely large, we get a Sperner's triangle $(\alpha_0 , \alpha_1 , \alpha_2)$ with $\psi_n(\alpha_i) = i$ . Then $\alpha_0 \sim \alpha_1 \sim \alpha_2$ (the triangle has infinitesimal size $\frac{1}{2^n}$ ) and by compactness of $\Delta$ and continuity of $f$ , there is a shadow $a = {}^\circ\alpha_0 = {}^\circ\alpha_1 = {}^\circ\alpha_2$ with $d(f(a) , a_i) \geq d(a , a_i)$ , $i = 0 , 1 , 2$ . But this implies $f(a) = a$ .

In higher dimensions, the proof is analogous, provided you generalize Sperner's lemma.

4. Weierstrass approximation theorem, following Bernstein.

Let $f : [0,1] \longrightarrow \mathbb{R}$ be standard continuous. For every (standard) $\varepsilon > 0$, there is a (standard) polynomial $P$ such that $|f(x) - P(x)| < \varepsilon$ on $[0,1]$.

As usual, put "standard" away to get the classical statement. But with the "standards", we have the equivalent formulation within I.S.T.

There is a polynomial $P$ such that, for every $x \in [0,1]$, $f(x) \sim P(x)$.

Of course, if $f$ itself is not a polynomial, $P$ is not standard and may have an infinitely large degree.

Proof. Fix two infinitely large integers $m$ and $n$ such that $\frac{m}{\sqrt{n}} \sim 0$ (for instance, take $n = m^4$). Call $I_x$ the set of all integers contained in $[nx - m\sqrt{n}\ ,\ nx + m\sqrt{n}]$ and $I'_x$ the set of integers in $[0, n]$ not contained in $I_x$.

Following Bienaymé Tchebicheff's inequality, we have

$$\sum_{q \in I_x} C_n^q (1-x)^{n-q} x^q \sim 1 \quad \text{and} \quad \sum_{q \in I'_x} C_n^q (1-x)^{n-q} x^q \sim 0 .$$

Consider the Bernstein polynomial $P(x) = \sum_{q=0}^{n} C_n^q f(\frac{q}{n}) x^{n-q} (1-x)^q$. We have $P(x) = \sum_{q \in I_x} + \sum_{q \in I'_x}$ and the second term is $\sim 0$, for $f(\frac{q}{n})$ is bounded by the standard number $\sup_{x \in [0,1]} |f(x)|$. Moreover, if $q \in I_x$, $f(\frac{q}{n}) = f(x + \frac{q}{n} - x)$ $= f(x) + \alpha_q$ with $\alpha_q \sim 0$, for $f$ is continuous and $|\frac{q}{n} - x| \leq \frac{m}{\sqrt{n}} \sim 0$.
Thus, for every $x \in [0,1]$, we get :

$$P(x) \sim \sum_{q \in I_x} C_n^q f(\tfrac{q}{n}) x^{n-q} (1-x)^q$$

$$\sim f(x)\Big(\sum_{q \in I_x} C_n^q x^{n-q}(1-x)^q\Big) + \sum_{q \in I_x} \alpha_q C_n^q x^{n-q}(1-x)^q \sim f(x) .$$

Note : this proof extends easily to $n$ variables.

5. Periodic orbits of some continuous vector fields in a closed tube.

Let $T = D^n \times S^1$ be a closed tube ( $D^n$ is the closed unit disk in $\mathbb{R}^n$ ), and $X$ a continuous vector field on $T$, pointing strictly inward along the boundary $\partial D^n \times S^1$, such that $b = d\theta(X) > 0$ on $T$ ( $\theta$ is the parameter along $S^1$ ). Then $X$ has at least one periodic orbit in $T$, with period $\tau \leq \frac{2\pi}{\inf b}$.

First look at some pictures which make the assumptions clear :

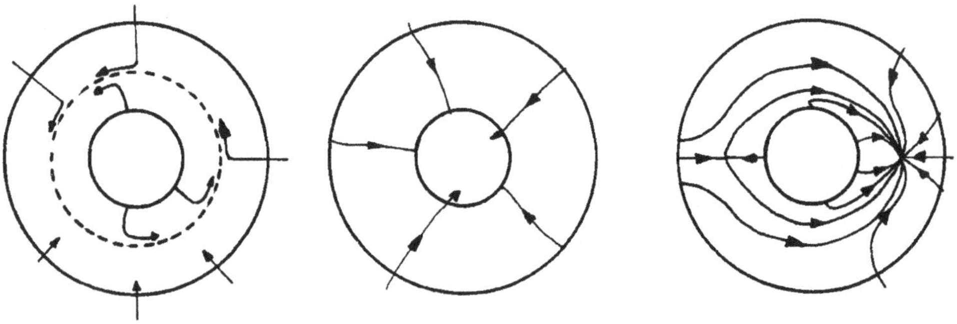

a nice case , with a périodic orbit .

not pointing inward along the boundary ; no periodic orbit .

pointing inward , but b changes sign ; two singular points , no periodic orbit .

Proof. a) Suppose that $X$ is of class $C^1$. Then, uniqueness of integral curves through an initial point and continuity yield a continuous map

$$\varphi : D^n \times \{\theta_0\} \longrightarrow D^n \times \{\theta_0\} ,$$

where $\varphi(x)$ is the first point where the orbit of $x$ meets again $D^n \times \{\theta_0\}$.

From Brouwer's fixed point theorem we get a point $x_0$ with $x_0 = \varphi(x_0)$, that is a periodic orbit of period $\tau$ such that $\int_0^\tau b\,dt = 2\pi$ ; thus $\tau \leq \frac{2\pi}{\inf b}$.

This is classical differential topology. Now, what is going on if

(b) $X$ is only continuous. In this case $\varphi$ is not defined and Brouwer cannot help us. We could try some sequence construction in order to use Ascoli's lemma. But why not a shadow-trick ?

Indeed, by transfer we may assume $x, T, X$ standard. From Weierstrass's approximation lemma, we get a $C^\infty$ vector field $Y$ on $T$ such that $Y(x) \sim X(x)$ at every $x \in T$. As $\partial T$ is compact, $Y$ is also strictly inward pointing along it (compare the radial components) and $\inf_T d\theta(Y) \sim \inf_T d\theta(X)$; this last number is standard $> 0$, hence $\inf_T d\theta(Y) > 0$. By (a) there is a periodic orbit $\gamma$ for $Y$ with finite period $\tau < \frac{2\pi}{\inf d\theta(Y)}$; for $t$ standard we have, for the shadow $\tilde{\gamma} = {}^\circ\gamma$,

$$\int_0^t X_{\tilde{\gamma}(u)}\, du \sim \int_0^t X_{\gamma(u)}\, du \sim \int_0^t Y_{\gamma(u)}\, du = \gamma(t) - \gamma(0) \sim \tilde{\gamma}(t) - \tilde{\gamma}(0) .$$

(use continuity of $X$ and of the standard operator $\int$.)

As both ends are standard, we get $\int_0^t X_{\tilde{\gamma}(u)}\, du = \tilde{\gamma}(t) - \tilde{\gamma}(0)$, which proves that $\tilde{\gamma}$ is an orbit of $X$; moreover, as $\tau$ is finite, we have

$$\tilde{\gamma}(t + {}^\circ\tau) \sim \tilde{\gamma}(t+\tau) \sim \gamma(t+\tau) = \gamma(t) \sim \tilde{\gamma}(t) ;$$

thus $\tilde{\gamma}$ is periodic of period ${}^\circ\tau \leq \frac{2\pi}{\inf d\theta(X)}$.

Remark. This is not the only case where properties of continuous fields can be deduced easily from the differentiable case.

## 6. Some remarks on diffeomorphism germs.

a) PROPOSITION 1. Two standard mappings $f, g : \mathbb{R}^n \longrightarrow \mathbb{R}^p$ ($n, p$ standard) have the same germs $\bar{f}$ and $\bar{g}$ at $0$ iff they are equal on the halo of $0$.

Proof. If $f$ and $g$ are equal on $h(0)$, they are equal on every open ball $B(0, \rho)$ with radius $\rho \sim 0$. This internal property is permanent until some standard radius, hence $\bar{f} = \bar{g}$. Converse obvious.

COROLLARY. $\bar{f} = 0$ iff $x \sim 0$ implies $f(x) \sim 0$.

b) Every standard germ may be represented by a standard mapping (transfer). We assume further that we write only standard representations.

c) Let $G$ be a standard group of $C^1$-diffeomorphism germs at $0$ in $\mathbb{R}^n$ ( $n$ standard). Assume that their tangent mapping at $0$ is the identity. If $u \sim 0$ in $\mathbb{R}^n$, and $t \in \mathbb{R}$, $t \neq 0$, we get an external mapping $\varphi^{(u,t)} : G_s \longrightarrow \mathbb{R}^n$ such that $\varphi^{(u,t)}(\overline{f}) = \frac{f(u) - u}{t}$.
This is well-defined for $f$ standard by proposition 1.

LEMMA: <u>If</u> $\overline{f}, \overline{g} \in G_s$, <u>one has</u> $\varphi^{(u,t)}(\overline{g} \circ \overline{f}) = \varphi^{(u,t)}(\overline{g}) + \varphi^{(u,t)}(\overline{f}) + \varepsilon\, \varphi^{(u,t)}(\overline{f})$ <u>with</u> $\varepsilon \sim 0$.

<u>Proof</u>. Use the mean value formula for each component to get

$$\varphi(\overline{g} \circ \overline{f})^i = \frac{1}{t}(g(f(u))^i - g(u)^i + g(u)^i - u^i) = Dg^i_v(\varphi(\overline{f})) + \varphi(\overline{g})$$

with $v = u + \theta_i(f(u) - u)$ and $0 < \theta_i < 1$.

As $f(0) = 0$, we get $0 \sim f(u) \sim u$ and $v \sim u \sim 0$; hence $Dg^i_v(\varphi(\overline{f})) - Dg^i_0(\varphi(\overline{f})) = \varepsilon_i\, \varphi(\overline{f})$ with $\varepsilon_i \sim 0$. But $Dg_0 = \mathrm{id.}$, which proves the lemma.

PROPOSITION 2. <u>If</u> $\varphi^{(u,t)}(\overline{f})$ <u>is finite for every</u> $\overline{f} \in G_s$, <u>the shadow</u> $H^{(u,t)}$ <u>of</u> $\varphi^{(u,t)}$ <u>is a (standard) morphism of</u> $G$ <u>into the additive group</u> $\mathbb{R}^n$.

This is an immediate consequence of the lemma (recall that $H$ is the unique standard mapping such that on standard $\overline{f}$, $H(\overline{f}) = {}^\circ(\varphi(\overline{f}))$, and that we have to prove $H(\overline{g} \circ \overline{f}) = H(\overline{g}) + H(\overline{f})$ on standard elements only.).

d) Whenever $\ell$ is not infinitesimal, proposition 2 applies, but $H$ is trivial ! Here is a case where $H$ is not trivial. Suppose that $G$ has a finite standard family of generators $\overline{a}_1, \ldots, \overline{a}_k, \overline{a}_1^{-1}, \overline{a}_k^{-1}$. Following a), there is an $u \sim 0$ such that $a_1(u) \neq u$ (for $a_1 \neq$ identity) ; hence

$$t = \sup_i\{\|a_i(u) - u\|, \|a_i^{-1}(u) - u\|\} > 0 .$$

As every standard $\overline{f}$ in $G$ is a finite product of generators, we get from the lemma $\varphi^{(u,t)}(\overline{f})$ finite ; thus proposition 2 works and as $\|H^{(u,t)}(\overline{a}_i)\| = 1$ for some $i$, $H^{(u,t)}$ is not trivial.

e) By transfer, a lemma used by THURSTON (" A generalization of the Reeb stability theorem" in Topology 13 (1974) to solve stability problems in foliation theory ; the original proof was

LEMMA. Any non trivial group of diffeomorphism germs at 0 in $\mathbb{R}^n$ finitely generated by germs which are tangent to the identity admits a non trivial group morphism on $\mathbb{R}^n$ .

The proof above, suggested by Reeb in 1976 has an historical meaning for us ; it was the starting point of our interest (and of some geometrically minded friends) for N.S.A.

Notice that, after a fascinating debate, J.P. JOUANOLOU produced a classical short proof, using Nakayama's lemma (see Topology 17, n° 1 (1978)).

Indeed, the nonstandard proof has a little advantage, for it is quite natural (and everybody knows the mean value formula...).

---

## Lesson 7

## INTEGRAL CURVES OF VECTOR FIELDS ON $\mathbb{R}^p$

THEOREM 1. Let $X$ be a continuous bounded vector field on $\mathbb{R}^p$ . For every $a \in \mathbb{R}^p$, there is at least one integral curve $\gamma$ of $X$ , such that $\gamma(0) = a$ .

THEOREM 2. If $X$ is lipschitzian, the integral curve through $a$ is unique. Note it $\gamma(a, t)$ .

THEOREM 3. If $X$ is lipschitzian, the flow $\gamma(a, t)$ is continuous and $\gamma(\gamma(a, t), s) = \gamma(a, t+s)$ for every $a, t, s$ .

THEOREM 4. If $X$ and $Y$ are lipschitzian and if $\|X - Y\| < K$ , their respective flows $\gamma$ and $\delta$ satisfy $\|\gamma(a, t) - \delta(a, t)\| \leq K|t|$ for every $a, t$ .

THEOREM 5. Under quite general assumptions, the shadow of an integral curve is an integral curve of the shadow.

<u>Comments</u>. 0) This lesson has to do with an essential pillar of differential wisdom. The reader may easily restrict the results in local form, in case $X$ is only defined and bounded on a neighbourhood of the starting point $a$ .

1) Whatever proof of theorem 1 you consider, there is a combinatorial procedure to get an approximate solution and then a fight with the epsilon's. In a non standard proof too, there is such a preliminary step (the "geometric idea"), but the second part is a simple shadow-trick.

For esthetical grounds, we use P. Harthmann's delay integrating technic rather than the usual Euler polygons, but the main trick is suitable for both approaches (see "Ordinary differential equations" New-York).

2) <u>Proof of theorem 1</u>. By transfer, we may assume all constants in the problem standard (i.e. $p$, $X$, $a$, a bound $M$ of $X$ ) and we expect a standard integral curve $\gamma$ . We restrict the proof to positive times and complete $\gamma$ by time reversal.

As integration is a standard operator, we look for a continuous standard mapping $\gamma : [0,\infty[ \longrightarrow \mathbb{R}^p$ satisfying system

$$(1) \qquad \begin{cases} \gamma(0) = a \\ \gamma(t) = \gamma(0) + \int_0^t X_{\gamma(u)}\, du \quad \text{for every standard } t \in [0,\infty[ \, . \end{cases}$$

Instead of (1), we solve by finite induction system

$$(2) \qquad \begin{cases} \alpha(t) = \lambda(t) \quad \text{on } [0,\tau] \\ \alpha(t) = \lambda(\tau) + \int_\tau^t X_{\alpha(u-\tau)}\, du \quad \text{on } [\tau,\infty[ \, , \end{cases}$$

where $\tau$ is a fixed positive infinitesimal and $\lambda : [0,\tau] \longrightarrow \mathbb{R}^p$ some given continuous curve such that $\|\lambda(t) - \lambda(0)\| \leq tM$.

We get an unique solution $\alpha$ satisfying (3) $\|\alpha(t+h) - \alpha(t)\| \leq hM$ for all $t$ , $t+h > 0$ ; this ends the combinatorial preliminaries.

Now choose $\lambda(0) \sim a$ and from (3) you get $\|\alpha(t)\| \leq \lambda(0) + tM$ , hence $\alpha(t)$ finite for every standard $t$ ; thus $\alpha$ has a shadow $\gamma = {}^\circ\alpha$ with

$\|\gamma(t+h)-\gamma(t)\| \leq hM$ for $t$, $h$ standard. This $\gamma$ is therefore continuous and for every $u \in [0,t]$, $t$ standard, we have $\gamma(u) \sim \gamma(°u) \sim \alpha(°u) \sim \alpha(u-\tau)$ and, due to the continuity of $X$, $X_{\gamma(u)} \sim X_{\gamma(°u)} \sim X_{\alpha(u-\tau)}$. As the standard operator $\int$ is continuous, this implies

$$\int_0^t X_{\gamma(u)}\, du \sim \int_\tau^t X_{\gamma(u)}\, du \sim \int_\tau^t X(\gamma(u-\tau))\, du = \alpha(t) - \lambda(\tau) \sim \gamma(t) - a .$$

As both ends are standard, we get

$$\begin{cases} \gamma(0) = a \\ \gamma(t) = a + \int_0^t X_{\gamma(u)}\, du , \end{cases}$$

which ends the proof.

3) The simplest $\lambda$ we may put in our "non standard computer" is the constant $t > a$. But different choices may give other integral curves, in case of non uniqueness. For instance, consider equation $\dot{\gamma} = \sqrt{\gamma}$ on $\mathbb{R}$ ; for $\lambda = 0$ you get the trivial solution $\gamma = 0$ through the origin. But for $\lambda(t) = \tau$ on $[0,\tau]$, you get another one (minorate $\alpha(n\tau)$ by $2^{n-1}\tau$ ).

4) <u>Proof of theorem 2</u>. Consider a field $X$ with $\|X_x - X_y\| \leq K\|x-y\|$ for some standard $K$, and two (standard) integral curves $\gamma_1$, $\gamma_2$ starting at the standard point $a$ : Call $\Delta_n = \sup_{n\tau \leq u \leq (n+1)\tau} \|\gamma_1(u) - \gamma_2(u)\|$ ; from equation (1) above, we get $\Delta_n \leq \Delta_{n-1}(1+\tau K)$, hence $\Delta_n \leq (1+\tau K)^n \Delta_0$. But $\Delta_0 \sim 0$ and $(1+\tau K)^n \sim \exp Kn\tau$ ; thus, for $n\tau$ finite, we have $\Delta_n \sim 0$ and, as every standard $t$ is contained in some internal $[n\tau, (n+1)\tau[$ with $n\tau$ finite, this proves that $\gamma_1(t) \sim \gamma_2(t)$. But both are standard, hence equal.

5) <u>Exercises</u>. • Prove theorem 3 and 4 along the same lines.

• Starting with the definition of the Lie bracket $[X,Y](f) = X(Y(f)) - Y(X(f))$, prove that for $X$, $Y$, $a$ standard and $\tau \sim 0$ (of course, $X$ and $Y$ have to be smooth enough), one has

$$[X,Y]_a \sim \frac{1}{\tau^2}(\tilde{a} - a) \quad \text{where } \tilde{a} \text{ is defined as follows :}$$

$$\begin{cases} a+\tau X_a = b \\ b+\tau Y_b = c \end{cases} \qquad \begin{cases} c-\tau X_c = d \\ d-\tau Y_d = \tilde{a} \end{cases}$$

6) A precise statement of theorem 5 is the following :

Let $X$ be a standard continuous vector field on a standard bounded open set $U$ and $Y$ a continuous vector field on $U$ such that for every $x \in U$, $X(x) \sim Y(x)$. Let $\gamma : [0, t_0] \longrightarrow U$ be some integral curve of $Y$, with $t_0$ standard. Then its shadow $\bar{\gamma}$ is an integral curve of $X$.

Proof. The shadow $\bar{\gamma}$ exists for every point of $U$ has a shadow. Then, for $t$ standard, $t \leq t_0$, we have

$$\int_0^t X_{\bar{\gamma}(u)}\, du \sim \int_0^t X_{\gamma(u)}\, du \sim \int_0^t Y_{\gamma(u)}\, du \sim \gamma(t) - \gamma(0) \sim \bar{\gamma}(t) - \bar{\gamma}(0) .$$

Both ends are standard, hence equal and $\bar{\gamma}$ is an integral curve of $X$. Notice that $X = {}^{\circ}Y$ needs only $X(x) \sim Y(x)$ for $x$ standard ; this would not be sufficient in the proof above. But the classical approximation lemmas mainly concern uniform approximations, which therefore satisfy our hypothesis. We gave in lesson III.6.§ 5 an application of theorem 5.

This also is another way to get integral curves of different kinds for fields without uniqueness property : approaching them by various smooth fields (say by Weierstrass's theorem), we get various flows which are "nearly integral" for the given fields (clearly, the shadow of such a flow is not usually a one-parameter group of diffeomrophisms.).

7) Exercise. Consider the two point boundary value problem

$$(P) \quad \begin{cases} \ddot{x}(t) + f(t, x(t), \dot{x}(t)) = 0 \\ x(a) = x(b) = 0 \end{cases}$$

Prove that Picard's iterative process (1893 !) gives a solution of (P), under suitable assumptions on $f$. Precisely, assume $f$ standard continuous with the

Lipschitz condition $|f(t,x,y)-f(t,\bar{x},\bar{y})| \leq K|x-\bar{x}| + L|y-\bar{y}|$ and consider the iterative process

$$\begin{cases} \ddot{x}_n(t) + f(t, x_{n-1}(t), \dot{x}_{n-1}(t)) = 0 \\ x_n(a) = x_n(b) = 0, \quad x_0(t) \text{ arbitrary.} \end{cases}$$

Prove that, provided $\frac{K(b-a)^2}{8} + \frac{L(b-a)}{2} < 1$, $x_n(t)$ has a shadow $x(t)$ which is a solution of (P) . Prove its uniqueness. (Hint : use the integral formulation

$$\begin{cases} x(t) = \int_a^b G(t,s)\, f(s, x(s), \dot{x}(s))\, ds \\ \dot{x}(t) = \int_a^b H(t,s)\, f(s, x(s), \dot{x}(s))\, ds \end{cases}$$

where

$$G(t,s) = \begin{cases} \frac{(b-t)(s-a)}{b-a} & \text{if } a \leq s \leq t \leq b \\ \frac{(b-s)(t-a)}{b-a} & \text{if } a \leq t \leq s \leq b \end{cases}$$

and $H = \frac{\partial G}{\partial t}$ . )

## Lesson 8

## THE INVERSE FUNCTION THEOREM

Theorem. Let $f : \mathbb{R}^p \longrightarrow \mathbb{R}^p$ be a $C^1$-mapping in a neighbourhood of some point $a$. If $Df_a$ is non singular, there exist open sets $U$ and $V$ with $a \in U$, $f(a) \in V$ such that the restriction of $f$ to $U$ is a $C^1$-diffeomorphism onto $V$.

Comments. 0) This is the second pillar of differential wisdom. We take as geometric idea for our proof the fact that the expected inverse $g$ of $f$ is "the envelope of its derivative $Dg$" ; as we expect that $Dg_{f(x)} = (Df_x)^{-1}$, this suggests the iterative process we shall use. This is an integration process and it is clear that non-standard analysis should help.

1) Recall from lesson I.5 that a standard $C^1$ mapping in a neighbourhood of a standard point $a$ satisfies the following property (apply the mean value formula to each component).

$$\forall\, x \sim a \,,\ \forall\, h \sim 0 \,,\ h \neq 0 \,,\quad \frac{f(x+h) - f(x)}{\|h\|} \sim Df_a\left({}^\circ\left(\frac{h}{\|h\|}\right)\right) \sim Df_x\left(\frac{h}{\|h\|}\right) .$$

2) Proof. By transfer, we may assume $p$, $f$, $a$, $U$, $V$ standard.

i) $f$ is one-to-one on some standard neighbourhood of $a$. Let $B(a,\eta)$ be an open ball with center $a$, radius $\eta$. If $\eta \sim 0$, $x, y \in B$, $x \neq y$, we have :

$\dfrac{f(y) - f(x)}{\|y - x\|} \sim Df_a\left({}^\circ\left(\dfrac{y-x}{\|y-x\|}\right)\right)$. As $Df_a$ is invertible, the second member is standard non zero ; hence $f(y) \neq f(x)$. By transfer, there is a standard $\eta$ with the same property (permanence principle also works). Let $\eta_0$ be such a number.

ii) $f$ is locally onto an open set. By continuity of $Df$ and compactness of the unit sphere, there is a standard $\rho > 0$, $\rho \leq \eta_0$ and a standard $m > 0$ such that for $\|u\| = 1$ and $x \in B(a, \rho)$, $\|Df_x(u)\| \geq m$.

Consider a standard point $c$ in the standard open ball $B(f(a), \rho m) = V$. Fix an infinitely large integer $\omega$ and define a mapping $\Phi : \mathbb{R}^p \longrightarrow \mathbb{R}^p$ by

$$\begin{cases} \Phi(x) = x + Df_x^{-1}(\dfrac{c - f(a)}{\omega}) & \text{on } B(a, \rho) \\[2mm] \Phi(x) = x & \text{out of } B(a, \rho) . \end{cases}$$

We have $\|\Phi(x) - x\| \le \dfrac{\|c - f(a)\|}{m\omega} \sim 0$ and, by induction,

$$\|\Phi^i(a) - a\| \le \dfrac{i}{\omega}\ \dfrac{\|c - f(a)\|}{m} < \dfrac{i\rho m}{\omega m} < \rho \quad \text{for } i < \omega ;$$

hence, all $x_i = \Phi^i(a)$ are in the ball $B(a, \rho)$ for $i < \omega$ and therefore $Df_{x_i}(x_{i+1} - x_i) = \dfrac{c - f(a)}{\omega}$ : We conclude that $x_\omega$ is finite and, if $°(x_\omega) = b$, we have $\|b - a\| \le \dfrac{\|c - f(a)\|}{m}$ (1) .



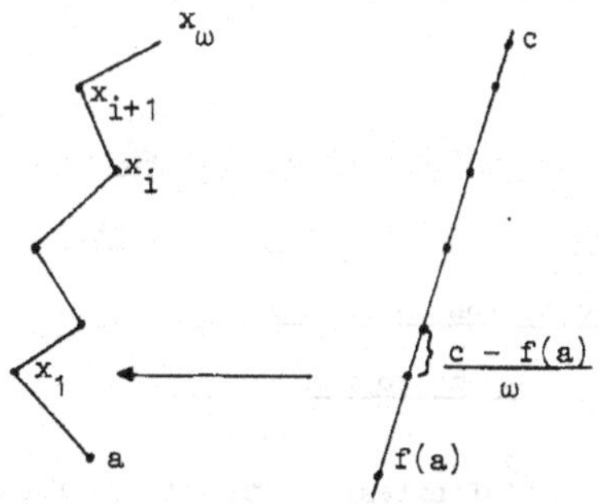

Then $f(b) \sim f(x_\omega)$

$$= f(a) + \sum_0^{\omega-1} f(x_{i+1}) - f(x_i)$$

$$= f(a)$$
$$+ \sum_0^{\omega-1} [Df_{x_i}(x_{i+1} - x_i) + \|x_{i+1} - x_i\|\varepsilon_i]$$

with $\varepsilon_i \sim 0$, because $x_{i+1} \sim x_i$ .

Thus $f(b) \sim f(a) + c - f(a) + r$ ,

with $\|r\| < \dfrac{\rho}{m} \dfrac{\sum \varepsilon_i}{\omega} \sim 0$ .

As both $f(b)$ and $c$ are standard, we get $f(b) = c$ ; we have proved that $f : U \longrightarrow V$ is one-to-one from $U = f^{-1}(V)$ onto $V$ .

iii) $g = f^{-1} : V \longrightarrow U$ _is_ $c^1$. From (1) we have $\|g(c) - a\| \le \dfrac{\|c - f(a)\|}{m}$ for every standard $c \in V$. By transfer, $y \sim f(a)$ implies $\dfrac{\|f^{-1}(y) - a\|}{\|y - f(a)\|}$ finite. This is true for every standard $a \in U$. Hence, if $k \sim 0$ ,

$$\Delta = Df_a(g(f(a) + k)) - a - Df_a^{-1}(k) = \|h\|\ \varepsilon$$

with $\varepsilon \sim 0$ and $h = g(f(a) + k) - a$ . As $\dfrac{\|h\|}{\|k\|}$ is finite, we have $\dfrac{\Delta}{\|k\|} \sim 0$ and, by continuity,

$$\dfrac{1}{\|k\|} [g(f(a) + k) - g(f(a)) - Df_a^{-1}(k)] \sim 0$$

which ends the proof, continuity of $Dg$ being a consequence of $Dg_y = Df_{g(y)}^{-1}$ .

3) This proof suggests that some problems which are usually solved by means of the inverse function theorem may be solved by a direct "shadow trick" (there are some examples in section IV). We had an analogous remark about Ascoli's lemma in lesson III.5.

---

## Lesson 9

## INFINITESIMAL TRANSFORMATIONS AND VECTOR FIELDS ON MANIFOLDS

THEOREM 1. Let $M$ be a standard compact topological space. Every infinitesimal transformation $(\xi, \bar{\xi}, \tau)$ on $M$ generates a standard continuous one parameter group of homeomorphisms of $M$.

THEOREM 2. Moreover, if $M$ is given a standard $C^1$-manifold structure, and if $X$ is a locally lipschitzian standard vector field on $M$ with respect to this structure, then there is an infinitesimal transformation whose one parameter group has $X$ as field of derivatives at $0$ with respect to the parameter.

Comments. 0) Some traditional geometers still call "infinitesimal transformations" what is nowadays called "vector fields", specially if the flows of these fields leave something invariant (e.g. a differential form, a geometric structure...) ; mysterious conditions like $\mathcal{L}_X \omega = 0$ (Lie derivative) make freshmen anxious as long as they have not clearly understood the relation between vector fields - sorry, sir, infinitesimal transformations... - and the one parameter groups they generate.

From an heuristic point of view, this relation is clear : for "little" parameters, replace the integral curves by tangent vectors, as physicists do. But the classical formal point of view is far from this idea, specially as regards vector fields on manifolds.

Moreover, what fails in classical mathematics is an intermedium between vector fields and flows, which really should be an "infinitesimal transformation" (that is, only displace the points a little) and make integration as natural as possible. It is easy to fill up this gap within the non standard frame.

1) As usual, we have to restrict the game to standard spaces (Recall once more that all well-known spaces are standard and that internal statements transfer to statements about standard objects).

Write $u \sim v$ for ${}^\circ u = {}^\circ v$ (the shadows exist, for $M$ is compact). Consider a pair $(\xi, \tau)$ where $\xi : M \longrightarrow M$ is an (internal) mapping such that for every $x \in M$, $\xi(x) \sim x$, and $\tau$ is a positive infinitesimal in $\mathbb{R}$ (consider $\tau$ as the time you need to compute $\xi(x)$. Such a pair generates a standard mapping $\gamma : M \times \mathbb{R}^+ \longrightarrow M$ (use the construction principle) which on standard arguments $(x, t)$ equals ${}^\circ(\xi^{[t/\tau]}(x))$, where $[\ ]$ is the "integral part". This "half-flow" has few general properties, apart from $\gamma(x, 0) = x$ ; therefore we make some restrictive assumptions, namely

i) <u>if</u> $n\tau \sim 0$ $(n \in \mathbb{N})$, <u>then</u> $\xi^n(x) \sim x$ <u>for every</u> $x \in M$.

ii) <u>if</u> $n\tau$ <u>is finite, and</u> $x \sim y$, <u>then</u> $\xi^n(x) \sim \xi^n(y)$.

With these conditions, we call $(\xi, \tau)$ a <u>semi-infinitesimal transformation</u>. To get also negative times, we consider a s.i.t. $(\bar{\xi}, \tau)$ with the same time-basis and the additional condition

iii) <u>if</u> $n\tau$ <u>is finite, then</u> $\xi^n \bar{\xi}^n(x) \sim \bar{\xi}^n \xi^n(x) \sim x$ <u>for every</u> $x \in M$.

We call such a triple $(\xi, \bar{\xi}, \tau)$ an <u>infinitesimal transformation</u> on $M$. Putting together both "half-flows" (with time reversed for $\bar{\xi}$), we get the <u>flow</u> $\gamma : M \times \mathbb{R} \longrightarrow M$ of $(\xi, \bar{\xi}, \tau)$.

2) Theorem 1 means that $\gamma$ is a continuous group of homeomorphisms. Note that for standard $t$, the mapping $\gamma(\cdot, t)$ is standard too. Suppose $x$, $t, s$ standard, $t \geq 0$, $s \geq 0$ and put $n = [\frac{t}{\tau}]$, $m = [\frac{s}{\tau}]$. Then $\gamma(x,t) \sim \xi^n(x)$ and $\gamma(\gamma(x,t), s) \sim \xi^m(\gamma(x, t))$. From conditions (i) and (ii), we get $\xi^m(\gamma(x,t)) \sim \xi^{m+n}(x) \sim \gamma(x, t+s)$ because $[\frac{t+s}{\tau}] = m+n$ or $m+n+1$. Hence $\gamma(\gamma(x,t), s) = \gamma(x, t+s)$.

Some proof with $\bar{\xi}$ for negative times.

From condition (iii), we get $\gamma(\gamma(x,-t),t) \sim \xi^n(\gamma(x,-t)) \sim \xi^n\bar{\xi}^{-n}(x) \sim x$ which implies $\gamma(\gamma(x,-t),t) = x$ (both ends are standard) ; same proof for $\gamma(\gamma(x,t),-t) = x$ . Thus $\gamma(\gamma(x,t),s) = \gamma(x,t+s)$ for $x,t,s$ standard, hence by transfer, for all $x,t,s$ .

Furthermore, conditions (i) and (ii) make the mappings $(x,t) \longrightarrow \xi^{[t/\tau]}(x)$ and $(x,t) \longrightarrow \bar{\xi}^{[t/\tau]}(x)$ s-continuous at every standard point. Therefore $\gamma$ is a continuous group of homeomorphisms (see lesson III.4).

3) Two i.t. $(\xi,\bar{\xi},\tau)$ and $(\xi',\bar{\xi}',\tau')$ are called <u>equivalent</u> if they generate the same flow. It is easy to check that another condition for equivalence is

iv) <u>if</u> $n\tau$ <u>is finite and</u> $n'\tau' \sim n\tau$ , <u>then</u> $\xi^n(x) \sim \xi'^{n'}(x)$ <u>and</u> $\bar{\xi}^n(x) \sim \bar{\xi}'^{n'}(x')$ .

For instance, if $\gamma$ is the flow of $(\xi,\bar{\xi},\tau)$ , then $(\gamma_\tau,\gamma_{-\tau},\tau)$ is an i:t: which is equivalent to $(\xi,\bar{\xi},\tau)$ (here $\gamma_\tau(x) = \gamma(x,\tau)$) :

4) If $f : M \longrightarrow N$ is a standard homeomorphism, and $(\xi,\bar{\xi},\tau)$ an i.t. on $M$ , then $(h\circ\xi\circ h^{-1}, h\circ\bar{\xi}\circ h^{-1},\tau)$ is an i.t. on $N$ and the respective flows are $\gamma_t$ and $h\circ\gamma_t\circ h^{-1}$ . This external mapping between both external sets of i.t. is clearly one-to-one.

5) Suppose that $\{(U,r)\}$ is a standard finite atlas on $M$ , with $C^1$ transition mappings. Theorem 2 relates some equivalence classes of i.t. (which are topological objects) with standard vector fields associated with the given atlas.

<u>Proof of theorem 2</u>. The field $X$ is given by standard lipschitzian fields $X^U$ on each $r(U) \subset \mathbb{R}^p$ ( $p$ is standard, of course). It is easy to get from $\{(U,r)\}$ a finer standard finite atlas $\{(V_i,r_i)\}_{i\in I}$ , such that :

- for every $i \in I$ , $r_i(V_i)$ is an open ball of standard radius $2\rho_i > 0$ ;
- the open sets $W_i = r_i^{-1}$ (open ball fo radius $\rho_i$ , same center as $r_i(V_i)$ cover $M$ .

On $r_i(V_i)$ , $X^i$ is $k_i$-lipschitz, with $k_i$ standard.

Put $k = \sup_{i\in I} k_i$ , $\rho = \inf_{i\in I} \rho_i$ , $\mu = \sup_{i\in I}(\sup_{x\in V_i} \|X^i_{r_i(x)}\|)$ ; these numbers are standard $> 0$ , if we exclude the trivial case $X = 0$ , whose solution is $\xi(x) = \bar{\xi}(x) = x$ .

As the $W_i$ cover $M$ , there is a standard map $\lambda : M \longrightarrow I$ such that for every $x \in M$ , $x \in W_{\lambda(x)}$ (use the construction principle, and prove the property on the standard $x$ ).

Note $(V_x, W_x, r_x)$ instead $(V_{\lambda(x)}, W_{\lambda(x)}, r_{\lambda(x)})$ and $X^x$ for $X^{r_x(V_x)}$ . Then we define an internal mapping $\Phi : M \times ]-\frac{\rho}{\mu}, +\frac{\rho}{\mu}[ \longrightarrow M$ by $\Phi(x, t) = r_x^{-1}(r_x(x) + t X^x_{r_x(x)})$ ; as $X^x$ is standard bounded and $r_x$ standard continuous, we have $\Phi(x, t) \sim x$ for every $t \sim 0$ . Choose a $\tau \sim 0$ , $\tau > 0$ and put $\xi(x) = \Phi(x, \tau)$ , $\bar{\xi}(x) = \Phi(x, -\tau)$ . We claim that $(\xi, \bar{\xi}, \tau)$ is an i.t. on $M$ . (Notice that $\Phi$ is precisely what intuition suggests as "infinitesimal flow" of $X$ .)

LEMMA. <u>If</u> $x \in W_z$ , <u>then</u> $r_z(\xi(x)) = r_z(x) + \tau X^z_{r_z(x)} + \tau\varepsilon$ <u>and</u> $r_z(\bar{\xi}(x)) = r_z(x) - \tau X^z_{r_z(x)} + \tau\bar{\varepsilon}$ <u>with</u> $\varepsilon \sim \bar{\varepsilon} \sim 0$ .

This follows immediately from the fact that $X^z$ and $X^x$ are related under the $c^1$-diffeomorphism $r_z \circ r_x^{-1}$ (use the characterization of " $c^1$ " in lesson I.5.)

Now, put $n_o = [\frac{\rho}{\mu\tau}]$ . A little computation shows that, for $n \leq n_o$ , $\xi^n(W_z) \subset V_z$ and yields the following inequalities

i) $\|r_z(\xi^n(x)) - r_z(x)\| \leq n\tau\mu$ ;

ii) $\|r_z(\xi^n(x)) - r_z(\xi^n(y))\| \leq \|r_n(x) - r_n(y)\|(1 + n\tau k)^n$ ;

iii) $\|r_z(\bar{\xi}^n\xi^n(x)) - r_z(x)\| \leq n\tau^2 k\mu$

and the corresponding relations for $\bar{\xi}$ . Thus, if $n\tau \sim 0$ , we have $n < n_o$ (because $n_o\tau \sim \frac{\rho}{\mu}$ is finite and not $\sim 0$ ) and from i), $\xi^n(x) \sim \bar{\xi}^n(x) \sim x$ .

If $x \sim y$ , use (ii) with $z = {}^\circ x = {}^\circ y$ to get $\xi^n(x) \sim \xi^n(y)$ for $n \leq n_o$ ; for other $n$ with $n\tau$ finite, $\frac{n}{n_o} = \frac{n\tau}{n_o\tau}$ is finite too ; put $\ell = [\frac{n}{n_o}]$ and you get $\xi^n(x) = \xi^{n_o\ell}(\xi^{n-n_o\ell}(x)) \sim \xi^{n_o\ell}(x)$ , because $(n - n_o\ell)\tau \sim 0$ . Use external induction (cf. Lesson I.6) on $\ell$ to infer that $\xi^n(x) \sim \xi^n(y)$ and $\overline{\xi}^n(x) \sim \overline{\xi}^n(y)$ whenever $x \sim y$.

In the same way, deduce from (iii) that $\overline{\xi}^n\xi^n(x) \sim \xi^n\overline{\xi}^n(x) \sim x$ for $n\tau$ finite. Thus $(\xi, \overline{\xi}, \tau)$ is an i.t. Call $\gamma$ its flow. (Note the importance of " $n\tau$ finite" which, in the proof above, makes $\ell$ standard ; indeed, external induction only works on standard integers). To finish the proof, we have to compute the derivative of the flow $\gamma$ with respect to $t$, at a standard point $x \in M$. In the chart $(V_x, r_x)$, put $\Delta = \|r_x(\xi^n(x)) - r_x(x) - tX^x_{r_x}(x)\|$ where $t$ is standard $> 0$ and $n = [\frac{t}{\tau}]$. If $t < \frac{\rho}{\mu}$, use (i) with $z = x$ to get $\Delta \leq (\tau n)^2 k\mu \sim t^2 k\mu$ ;

$$\text{hence } \frac{\Delta}{t} \leq tk\mu \text{, which proves that } \frac{1}{t}(r_x(\gamma(x, t) - r_x(x)) \xrightarrow[t \to 0]{} X^x_{r_x}(x) \text{,}$$

because $\gamma(x, t) \sim \xi^n(x)$. Same limit for $\overline{\xi}$. By transfer, we get the expected result.

6) This proof is quite a mouthful ; this is not surprising : vector fields on manifolds are rather artificial and complicated ; to get an intermedium which makes integration easy needs some work

As a reward, we get the classical completeness property :

<u>On a compact</u> $c^1$ <u>manifold, every locally lipschitz vector field</u> $X$ <u>generates a global continuous one parameter group of homeomorphisms.</u>

<u>Proof</u>. Put standard everywhere and consider the flow of an i.t. associated with $X$ as in theorem 2. If $X$ is $C^\infty$, use the classical proof to get $\gamma$ also $C^\infty$.

7) Let $(\xi, \overline{\xi}, \tau)$ be an i.t. on a standard manifold $M$ with a $C^1$ atlas $\{(U, r)\}$. If we try to get a standard vector field $X$ whose flow is the same as the flow of $(\xi, \overline{\xi}, \tau)$, we have to consider

$$X^U(u) = {}^\circ\!\left(\frac{(r \circ \xi \circ r^{-1})(u) - u}{\tau}\right) \text{ on } U \text{.}$$

If these shadows exist, the fields $X^U$ are compatible with chart transitions and we get a vector field $X$, locally lipschitzian in the good cases ; its flow is the same as the flow of $(\xi, \overline{\xi}, \tau)$ (that is the i.t. we get from $X$ as in theorem 2 is equivalent to the given one).

But the shadow need not exist and we have more equivalence classes of i.t. than standard vector fields for a given $c'$-atlas. Indeed, we have enough i.t. to get the vector fields of all possible $C^1$-structures on $M$ .

This is a new insight in global differential geometry : we may think that usual "tangent objects", like tensor fields, which are given as a "ready to glue kit" via an atlas, correspond to intrinsic topological objects (after restriction to standard spaces and standard fields), defined up to an external equivalence, for which integration and derivation are natural operations independent form any atlas.

The starting point is the fact that a pair of points $(x, y)$ such that $°x = °y$ on a standard topological space is an accurate intermedium between the intuitive idea of "little displacement" and the classical concept of tangent vector.

In lesson 10, we outline some ideas in this direction.

8) We considered only compact spaces here, to avoid trouble with local groups. However, the reader may easily define i.t. and their maximal flow in locally compact standard spaces.

---

## Lesson 10

## SOME INTERMEDIATE OBJECTS IN DIFFERENTIAL AND ALGEBRAIC TOPOLOGY

THEME 1. Let $M$ be a standard topological space. Define "predifferential" objects for which derivation and integration operations are outlined, and which correspond to the usual differential objects associated with a smooth manifold structure on $M$ .

THEME 2. Describe intermediate objects on standard spaces which clarify the relations between various cohomological theories.

<u>Comments</u>: 1) Both themes above set up an ambitious programme which goes far beyond the frame of this book. In lesson 9, we worked it out for vector fields. Along the same lines, we may consider differential forms, ... Here we only give some landmarks.

2) As an introduction to theme 1, recall J. Milnor's motivations for his Microbundles (see Topology 1964, vol. 3, suppl. 1, pp. 53-80) :
"...; Suppose that one tries to construct something like a "tangent bundle" for a manifold $M$ which has no differential structure. Each point $x \in M$ has neighbourhoods which are homeomorphic to Euclidean space. It would be plausible to choose one such neighbourhood $U_x$ for each $x$, and to call $(x) \times U_x$ the "fibre" over $x$. Unfortunately however, it seems difficult to choose such a neighbourhood $U_x$ simultaneously for each $x \in M$, in such a way that $U_x$ varies continuously with $x$. Furthermore even if such a choice were possible, it is not clear that the resulting object would be a topological invariant of $M$. To get around these difficulties we consider a new type of bundle, in which the fiber is only a "germ" of a topological space. Thus for the tangent microbundle of $M$, the fibre over $x$ is a completely arbitrary neighbourhood of $x$ (subject only to the uniformity condition that the set of all $(x, y)$ with $y \in U_x$ should form a neighbourhood of the diagonal in $M \times M$). At any stage of the argument we will be allowed to pass to smaller neighbourhoods ; hence any particular choice of the $U_x$ becomes irrelevant."

The non standard answer to these provocative lines is clear : consider (on a standard space $M$) the external set of all $(x, y) \in M \times M$ such that $°x = °y$.

3) This is a precise formulation of an heuristic guideline used by W.T. VAN EST and Th. J. KORTHAGEN in "Non enlargible Lie algebras" (see Indag. math. 26, n° 1, 1964). The question is about the isomorphism between Alexander and de Rham cohomology.
"Let $F$ be a real valued $C^\infty$ function defined on some neighbourhood of the dia-

gonal $D$ in $M^{n+1}$. For convenience, we assume for a moment that $F/D = 0$. Let $x_o, \dots, x_n$ be an $(n+1)$-tuple on $M$ and suppose that $x_1, \dots, x_n$ lie "very close" to $x_o$. The ordered pairs $x_o x_i$ are "pratically" tangent-vectors at $x_o$, and thus $F$ defines for any $x_o \in M$ a function $(F)$ on the $n$-tuples of tangent vectors at $x_o$. Applying the alternation operator to $(F)$, it becomes an alternating function whose "main part" $\tau F$ is $n$-linear because of the differentiability of $F$. $\tau F$ is thus a differential form of degree $n$ on $M$ and is called the residue of $F$. From the construction it is clear that two functions $F$ that coincide on some neighbourhood of $D$ have the same residue. And we have $F \longrightarrow \tau F$ establishes an isomorphism of the Kolmogorov-Alexander-Spanier cohomology of $M$ with the de Rham cohomology."

This programme can be taken to the letter within N.S.A., provided $M$ and $F$ are standard ; recall from Lesson I.5 that taking the "main part" is an easy shadow trick.

4) Precisely, the procedure has two steps :

a) from standard function $F : M^{p+1} \longrightarrow \mathbb{R}$, define a function $\overline{F}$ by

$$\overline{F}(x_o, x_1, \dots, x_p) = \Sigma\, (-1)^i\, \mathcal{G}_p\, F(x_o, x_1, \dots, \hat{x}_i, \dots, x_p)$$

(where $\mathcal{G}_p\, F(x_o, u_1, \dots, u_p) = \underset{\sigma \in \mathfrak{S}_p}{S_g}\, F(x_o, u_{\sigma(1)}, \dots, u_{\sigma(p)})$.

This $\overline{F}$ is skewsymmetric on $(x_1, \dots, x_p)$.

b) On every chart of a standard finite atlas $(U, r)$, define $\omega_{r(x_o)}(v_1, \dots, v_p) = {}^{\circ}\!\left(\dfrac{\overline{F}(x_o, x_1, \dots, x_p)}{\varepsilon^p}\right)$ with $x_i = r^{-1}(r(x_o) + \varepsilon v_i)$ for each $i > 0$ and $\varepsilon \sim 0$.

As $F$ is smooth, $\overline{F}$ is nearly $p$-linear (see lesson I.5) and $\omega_{r(x)}$ is a differential form on $r(U)$ which clearly agrees with transition between charts. Of course, the operator $\delta F(x_o, \dots, x_{p+1}) = \Sigma\, (-1)^i F(x_o, \dots, \hat{x}_i, \dots, x_{p+1})$ yields an operator $\Delta\overline{F}(x_o, \dots, x_{p+1}) = \mathcal{G}(\Sigma(-1)^i\, \overline{F}(x_o, \dots, \hat{x}_i, \dots, x_{p+1}))$ ; an easy computation shows that applying the procedure above to $\Delta\overline{F}$ gives the usual exterior differential $d\omega$.

5) Now, forget any differential structure ; the remarks above suggest to call "infinitesimal flux" of degree $p$ on a standard topological space $M$ any standard mapping $H : M^{p+1} \longrightarrow \mathbb{R}$ such that

i) if ${}^\circ x_o = {}^\circ x_1 = \dots = {}^\circ x_p$ , then $H(x_o, x_1, \dots, x_p) \sim 0$ .

ii) $H$ is skew symmetric on $x_1, \dots, x_p$ and if $x_i = x_o$ , $H = 0$ .

iii) additionnal conditions concerning integration.

As for "differentiation", use operator $\Delta$ defined in § 4 .
Integration needs a supplementary object (which replaces the time basis $\tau$ of infinitesimal transformations), for instance a finite simplicial complex $K$ of dimension $p$ (a non standard one, of course) such that all vertices in a same p-face have the same shadow (assume $M$ compact). We define

$$\int_K H = {}^\circ\Big(\sum_\sigma H(\sigma_o, \sigma_1, \dots, \sigma_p)\Big)$$

where $\sigma$ runs on the set of all p-faces in $K$ , conviniently oriented. This shadow only exists if conditions (iii) are strong enough. We could also write conditions on pairs $(H, K)$ to get the same integrals.

6) If a (standard) differential structure is given, procedure b) in § 4 yields differential forms from some infinitesimal flux (but not from any one, of course). On the other hand, we may get a flux from a differential form using a procedure of the same kind as for vector fields : consider a standard mapping $\lambda$ which localizes each point in a chart of a convenient atlas and define $H$ by $H(x_o, x_1, \dots, x_p) = \omega_{r(x_o)}(r(x_1) - r(x_o), \dots, r(x_p) - r(x_o))$ , (where $r(x) = r_{\lambda(x)}$ ) on a neighbourhood of the diagonal in $M^{p+1}$ ; then extend it to $M^{p+1}$ arbitrarily.

As for the complex $K$ , take an infinitely fine subdivision of some standard triangulation of $M$ , or relate it with an infinitely fine partition of unity associated with the given differential structure. Then $\int_K H$ is the integral of $\omega$ on $M$ in the usual sense.

7) Exercises.

a) Write and prove a "Stokes formula" for infinitesimal flux.

b) Extend our definitions to involve predifferential objects for tensor fields of all kinds ; define a "preriemannien geometry" with curvature, parallel displacement and geodesics (without any "linear" aspect).

c) Prove that the mapping $\tau$ defined in § 4 yields an isomorphism between the Kolmogorov-Alexander-Spanier and de Rahm cohomologies (construct directly an inverse, without using fine sheaves)...

8) Theme 2 enlarges the discussion to more general homological theories. The keyword is always "refinement" ; we may replace it by the use of non standard intermedia, e.g. infinitely fine triangulations, infinitely fine open coverings, infinitely fine singular simplexes, etc..., and by means of them describe the isomorphisms which relate under convenient assumptions Cech, K.A.S., de Rahm and singular theories. Maybe, this could help to get a deeper insight in algebraic topology including some recent aspects.

---

## Lesson 11

## HOLOMORPHIC FUNCTIONS

THEOREM. Let $U$ be an open set in the field $\mathbb{C}$ of complex numbers, with all its points finite and with ${}^{\circ}U$ open ; let $f : U \longrightarrow \mathbb{C}$ be an holomorphic function, taking only finite values. Then ${}^{\circ}f$ is holomorphic on ${}^{\circ}U$ and its n-th derivative is the shadow of $f^{(n)}$, for every standard $n$.

Comments. 0) Recall that $\mathbb{C}$ is standard and that the shadow ${}^{\circ}U$ is the standard subset whose standard points are the shadows of the points of $U$.

Clearly, $f$ must be internal, if you work in ISTE.

1) The corresponding statement for real analytic functions is false, for

any standard continuous function is the shadow of some polynomial. The theorem above, a nice contribution of A. Robinson (see [R], p. 155) is an original consequence of holomorphy. Its classical counterpart has to do with limits of sequences, as usual.

2) <u>Proof</u>. We first prove continuity of ${}^\circ f$ on ${}^\circ U$, i.e. that $z \sim a$ implies $f(z) \sim f(a)$ at every standard point $a \in {}^\circ U$. Indeed, the maximum $m$ of $f$ on a standard closed disk of radius $\rho > 0$, center $a$, contained in $U$ is taken at a point $b$ of the boundary. This $m$ is finite and, using Schwarz's lemma, we get $|f(z) - f(a)| \leq |z - a| \frac{2m}{\rho} \sim 0$ if $z \sim a$.

Now consider any standard closed disk $D \subset {}^\circ U$. As $D$ is compact and ${}^\circ f$ continuous, use the familiar trick $f(z) \sim f({}^\circ z) \sim ({}^\circ f)({}^\circ z) \sim ({}^\circ f)z$ to get $\int_{\partial D} ({}^\circ f)(z)\, dz \sim \int_{\partial D} f(z)\, dz = 0$. Thus ${}^\circ f$ is holomorphic on ${}^\circ U$.

Now use Cauchy's formula at any standard point $a$, and for $n$ standard :

$$2\pi i \; ({}^\circ f)^{(n)}(a) = \int_{\partial D} \frac{({}^\circ f)(z)\, dz}{(z-a)^{n+1}} \sim \int_{\partial D} \frac{f(z)\, dz}{(z-a)^{n+1}} \sim 2\pi i \; f^{(n)}(a) .$$

Hence $({}^\circ f)^{(n)} = {}^\circ(f^{(n)})$.

3) It is easy to extend the theorem to $p$ variables ($p$ standard). We leave the details to the reader.

Now consider a standard compact complex analytic $p$-dimensional manifold $M$ with a finite standard atlas $\{(U_i , r_i)\}$ and an holomorphic function $f : M \longrightarrow \mathbb{C}$. Then $f(M)$ is bounded but $\mu = \sup|f(M)|$ need not be finite ; however $\frac{f}{\mu}$ is finitely valued and we reduce the discussion to such functions. Then ${}^\circ f$ exists and, as the $U_i$ are standard, we may apply the extended theorem to the holomorphic functions $g_i = f \circ r_i^{-1} : r_i(U) \longrightarrow \mathbb{C}$ ; thus ${}^\circ g_i$ is holomorphic on $r_i(U)$ and, as $r_i$ is standard, $({}^\circ f) \circ r_i^{-1} = {}^\circ(f \circ r_i^{-1}) = {}^\circ g_i$, which proves that ${}^\circ f$ is holomorphic on $M$. (Standerdness of $M$ is essential here.)

4) Furthermore, if $N$ is another standard complex analytic manifold and $f : M \longrightarrow N$ an holomorphic mapping such that $f(M) \subset h(N)$ (the halo of $N$),

use a standard finite atlas of $N$ to prove that $^{\circ}f$ is holomorphic. For instance, consider an holomorphic vector field $X$ on $M$, that is a section $X : M \longrightarrow TM$ of the tangent bundle. As $M$ is compact, we may divide $X$ by some constant to get a finitely valued vector field (this makes sense via a standard atlas of $M$) ; then $X$ has a shadow $^{\circ}X$ which is holomorphic. Moreover, for two such fields, $[X , Y]$ has a shadow and $[^{\circ}X , ^{\circ}Y] = {}^{\circ}[X , Y]$, because on standard holomorphic functions, one has $^{\circ}(X(f)) = (^{\circ}X)(f)$, which is easy to check.

Analogous remark for differential forms.

5) Let us give a last application. Consider an holomorphic Lie group structure on $M$, that is a group product $\mu : M \times M \longrightarrow M$ and an inversion $\nu : M \longrightarrow M$, with unit $e$. Then $^{\circ}\mu$, $^{\circ}\nu$, $^{\circ}e$ exist ($M$ is compact) and are holomorphic. The equations about group structure, e.g. $\mu(x , \nu(x)) = \mu(\nu(x) , x) = e$ go through the shadows and we get a Lie group structure again, which need not be isomorphic with the given one.

This is a considerable refinement of Bolzano-Weierstrass lemma indeed. It means that any sequence of complex Lie group structures on a compact manifold has a limit point which is a Lie group structure.

6) Exercise. Extend this result to a Lie group action $G \times M \longrightarrow M$ which $G$ and $M$ standard.

PART IV : NON STANDARD ANALYSIS AS A TOOL IN PERTURBATION PROBLEMS

## Lesson 0

## INTRODUCTION AND CHECK-LIST OF THE TOOLS

In section II, we tried to justify N.S.A. by logician's arguments : classical mathematics remain valid but bridges are thrown over the entry of some deep fjords, by means of the material that the new language is able to describe.

Section III was concerned with some spectacular abridgments ; of course, it was about a well-polished part of classical mathematics and only a willing reader, ready to change his mind, could consider these examples as a convincing efficiency test for N.S.A.

In Section IV, however, we try to illustrate on some important problems of current research that N.S.A. certainly is an efficient tool for people working on perturbation problems. Our aim is to get some easy informations about subjects which usually are considered as cumbersome.

For instance, we are concerned with perturbations and deformations of Lie algebras, and don't want to compute any cohomology ; we are also concerned with singular perturbations in differential equations, and want to proceed without computing asymptotic developments especially whenever asymptotics fail. The latter subject is very important in applied mathematics, due to the fact that engineers often are faced with big effects of very little perturbations in physical systems.

How does a perturbation problem look like ? You have a fixed object $P_0$ and you have to compare it with an object $P$ near $P_0$ within some underlaying topological space. If $P$ is like $P_0$ as regards some important aspect, it is a regular perturbation problem. If $P_0$ is somewhat degenerate with respect to $P$, it is a singular perturbation problem ; often $P_0$ is simpler than $P$ and the main question is to relate the fine properties of $P$ with the rough properties of $P_0$.

Usually, one has to follow the behaviour of some "parasite" of $P$ as $P$ tends to $P_o$ (e.g. roots of polynomials, ideals of algebras, integral curves of differential equations, etc...). The classical mathematical language is not rich enough to formulate directly the intuitive concept of "a perturbation of $P_o$" ; instead, one has to describe the effects on parasites of "being near enough" : statements look like "if $P$ is near enough to $P_o$, then...". Thus formulations, and a fortiori proofs, may be heavy and far from intuitive arguments.

Now, within N.S.A., a perturbation problem has a transfered form in which you have a standard $P_o$ and you define a perturbation of $P_o$ as a $P$ which (in some standard topology) is infinitely near $P_o$ (in other words, which is in the halo of $P_o$). Of course, you immediately try to compare the eventual shadows of $P$'s parasites with those of its shadow $P_o$.

Nice formulations occur, and also close-to-intuition proofs ; for instance, a technical difficulty in classical proofs is to match together asymptotic behaviours in two different domains ; within N.S.A. we have a very simple permanence principle, which allows to "round the corners" between contiguous behaviours.

All this will be abundantly illustrated in the next fourteen lessons ; indeed, we use only a little (but essential) part of N.S.A., so little that we give below a check-list of the necessary tools that should make the text readable even if you are not familiar with sections I, II, III.

Clearly, the reader may find after lesson 15 that, indeed, there is here much more geometry than Non-standard analysis ! We have two arguments for our defence :

- our everyday business is geometry...
- N.S.A. is just what fails in classical treatments to handle geometric problems - as differential equations certainly are, for instance - with a geometric mind. This is so true that papers on singular perturbations are filled up with computations and hard analysis, but rarely with geometric ideas !

Tool bundle with instructions for use :

• Introduce the adjective "standard" in your mathematical language. Call internal the statements which don't use this adjective (i.e. the statements of classical mathematics), and external the new statements.

• Introduce the following principles :

- all axioms of set theory when applied to internal statements.

- the transfer principle : it applies to any internal statement $A(x, t_1, \dots, t_k)$, whose only free variables are $x, t_1, \dots, t_k$ (consider $t_1, \dots, t_k$ as parameters) and whose fixed ingredients are known to be standard. The principle says that

$$\forall^{St} t_1 \dots \forall^{St} t_k \; (\forall x \; A(x, t_1, \dots, t_k) \Longleftrightarrow \forall^{St} x \; A(x, t_1, \dots, t_k))$$

(read "for every standard $x$" for " $\forall^{St} x$ ").

Use : to prove the statement $\forall x \; A(x, t_1, \dots, t_k)$, we equivalently prove $\forall^{St} x \; A(x, t_1, \dots, k)$ provided $t_1, \dots, t_k$ and all fixed ingredients of $A$ are standard.

- the idealization principle (in its weak form) works for any standard binary relation $\rho$, whose domain is a standard set $E$ and which satisfies the idealizability property : for every finite standard subset $F \subset E$, there is a standard $v \in E$ related to every $u \in F$.

Then, there is a $v \in E$ related to every standard point of $E$.

- the standardisation principle applies to any external statement $C(z)$ with free variable $z$.

For every standard set $E$, there is one and only one standard set $F$, whose standard elements are the standard elements of $E$ which have property $C$.

• Use the following consequences of these principles :

- all what is, is standard : in other words, any object which can be constructed by classical set-theoretic operations from standard object is standard. For instance all well-defined classical sets like $\mathbb{N}$, $\mathbb{Q}$, $\mathbb{R}$, $\mathbb{C}$, $\mathbb{R}^3$ ... are standard. $\mathbb{R}^n$ is standard if $n$ is standard. The image of a standard $x$ under a standard mapping $f$ is standard, etc...

- $\mathbb{N}$ (and also $\mathbb{R}$ ) have infinitely large elements (i.e. larger than any standard one), since the order relation is idealizable.

Any non-standard integer is infinitely large.

Reals with infinitely large inverse (in absolute value) are called infinitesimals.

- Write $x \sim y$ for " $x - y$ infinitesimal". This extends to $\mathbb{R}^n$, for any standard $x$, by means of projections or equivalently of the standard norm $\sup|x_i|$. Call finite ("limited" would be better) all points whose norm is bounded by some standard number. Then

- a standard subset of $\mathbb{R}^n$ ( $n$ standard) is bounded if all its elements are finite. Similar criterion for standard mappings.

- $f : \mathbb{R}^n \longrightarrow \mathbb{R}^p$, with $n, p, f$ standard is continuous at a standard point $a$ iff $x \sim a$ implies $f(x) \sim f(a)$.

- Every finite point in $\mathbb{R}^n$ ( $n$ standard) is infinitely close to $(\sim)$ to some standard point, called its <u>shadow</u>.

- for every mapping $f : \mathbb{R}^n \longrightarrow \mathbb{R}^p$ ( $n, p$ standard) which takes only finite values, there is a unique standard mapping $°f : \mathbb{R}^n \longrightarrow \mathbb{R}^p$, called its <u>shadow</u>, such that for every standard $x \in \mathbb{R}^n$, $(°f)(x) = °(f(x))$. By transfer, the internal properties of $°f$ may be proved on standard arguments only.

- for every subset $A \subset \mathbb{R}^n$ ( $n$ standard), there is a unique subset $°A \subset \mathbb{R}^n$, called its <u>shadow</u>, whose standard elements are the shadows of the finite elements of $A$. Notice that $°A$ may be empty and don't confuse $°\{a\}$ and $\{°a\}$ : the second is not defined whenever the first is empty. The internal properties of $°A$ may be proved on its standard elements only.

- a standard subset $A$ of $\mathbb{R}^n$ ( $n$ standard) is compact (for the usual topology) iff every element of $A$ has a shadow in $A$. Example : an interval $[a, b] \subset \mathbb{R}$, with $a$ and $b$ standard.

- a standard set is finite iff all its elements are standard.

- the halo of a standard subset $A \subset \mathbb{R}^n$ ( $n$ standard) is not a set but a manner of speaking : we say " $x$ is in the halo of $A$ " instead of " $x$ has a shadow in $A$ ".

Some presentations of N.S.A. include halos in their sets. <u>But never apply mathematical constructions to halos without care</u>.

- every external statement about standard objects has a classical counterpart, that may be deduced through successive transfers. The prototype is continuity.

• Caution : $\{x \in V , P(x)\}$ is only defined if $P$ is an internal property (the subset axiom only works for internal properties).

For instance, in $\mathbb{R}$ , there is no subset of standard (resp. non standard, resp. finite, resp. infinitely large) elements, since such subsets would have an upper (or lower) bound, which is clearly not possible.

As a very important counterpart, we get the various forms of the

• Permanence principle. Its general form is given in II.7.12. We need the following particular cases :

- if an internal property $A(x)$ is true for every finite $x \in \mathbb{R}$ , it remains true until some infinitely large $x$ .

(Proof : the subset $\{x , \forall y < x , A(y)\}$ contains all finite reals, hence also infinitely large ones.)

Similar statement to get permanence from "infinitely large" to "finite".

- if a mapping $f : \mathbb{R}^+ \longrightarrow \mathbb{R}^n$ ( $n$ standard) is such that for every finite $x$ , $f(x) \sim 0$ , this property is permanent until some infinitely large $x$ .

(Proof : consider the internal property $\forall y < x , \|f(y)\| < \frac{1}{y}$ and apply the first principle.)

In its sequencial version, this second form is known as "Robinson's lemma".

Caution :

- if $g(x) \sim 0$ as long as $x$ is not $\sim 0$ , this property is permanent until some $x \sim 0$ (put $f(x) = g(\frac{1}{x})$ ).

- if $\|g(x)\|$ is infinitely large as long as $x$ is not $\sim 0$ , this property is permanent until some $x \sim 0$ (put $f(x) = \frac{1}{\|g(\frac{1}{x})\|}$ ).

- if $\|g(x)\|$ is infinitely large for every finite $x$ , this property is permanent until some infinitely large $x$ (put $f(x) = \frac{1}{\|g(x)\|}$ ).

but if $g(x) \sim 0$ (or $\frac{1}{g(x)} \sim 0$ ) as long as $x \sim 0$ (or $x$ infinitely large), there is no permanence. Counter example : $g(x) = x$ or $g(x) = \frac{1}{x}$ .

Thus, to apply the principle in this second form, recall that x has to be finite or not infinitesimal (i.e. $\frac{1}{x}$ finite) at the entry.

• To use all these principles without fear, you have to know that any internal statement which has a proof involving the adjective "standard" has also a classical proof.

Of course, the non-standard proof should be shorter...

# Lesson 1

## PERTURBATIONS OF ALGEBRAIC EQUATIONS

Problem: Let $P_0 \in \mathbb{C}_n[X]$ be a complex polynomial of degree $n$. Compare the roots of $P_0$ with those of $P = P_0 + H$, whenever $H$ has bounded degree and little coefficients.

Formulation within IST. If $P_0$ is standard (hence $n$ standard), compare the roots of $P_0$ with those of a perturbation $P = P_0 + H$ with standard degree $m$.

THEOREM 1 (regular perturbation). Assume that $m = n > 0$. Then every root $a_i$ of $P_0$ with order $r_i$ is the shadow of roots of $P$ with total order $r_i$. Moreover, for each root $a_i$, there exists a standard $\mathbb{C}$-linear mapping $L_i : \mathbb{C}[X] \longrightarrow \mathbb{C}$ only depending on $P_0$ such that for every root $\alpha$ of $P$ with shadow $a_i$, one has $(\alpha - a_i)^{r_i} = L_i(H) + \|H\|\varepsilon$ with $\varepsilon \sim 0$.

THEOREM 2 (singular perturbation). Assume that $m > n > 0$. Then $P$ has $n$ finite roots (counted with orders) whose shadows are the roots of $P_0$ as in theorem 1 and $m - n$ roots with infinitely large moduli.

THEOREM 3. Let $P_0$ be a standard polynomial in $k$ variables and $S_0$ the algebraic hypersurface of its zeros in $\mathbb{C}^k$. Let $P$ be a perturbation of $P_0$ with standard degree, and $S$ the corresponding surface. Then $S_0$ is the shadow of $S$.

Comments. 0) We are concerned here with very simple perturbation problems, of basic importance in a lot of stability questions ; however their classical treatments are not trivial and even the formulation of the results is not pleasant: Let us feel the problem's flavour on a particular case.
Consider $P = \varepsilon x^2 + 2x - 1$. Its roots are $\frac{1 \pm \sqrt{1+\varepsilon}}{\varepsilon}$ ; if $\varepsilon$ tends to $0$, one root has limit $\frac{1}{2}$ (the root of $P_0 = 2x - 1$), the other tends to infinity. Quite

easy, isn't it ? But what's going on with the roots of

$$P = \varepsilon(x^7 + x^4 + 1) + x^2(x-8) .$$

There is no nice formula to get the roots... However, theorem 2 asserts that for any $\varepsilon \sim 0$, one root is near $8$, two are near $0$ (or a double one), and four (distinct or not) have infinitely large moduli ; moreover, consider the derivative $x(7\varepsilon x^5 + 4\varepsilon x^3 + 3x - 16)$ of $P$. It has no root near $0$, other than $0$ (use again theorem 2) which is not a root of $P$. Hence the roots of $P$ near $0$ are distinct. As you see, in every particular case, we may obtain a quite precise answer.

1) We define a perturbation $P$ of a standard polynomial $P_o$ as a polynomial of degree $m$ (standard or not) such that $P_o = {}^\circ P$ as functions, i.e. for any standard $z \in \mathbb{C}$, $P_o(z) \sim P(z)$. If $m$ is standard, this is equivalent to the fact that the coefficient of $x^i$ in $P$ is infinitely near the corresponding coefficient of $P_o$ whenever $i \le n$ and near $0$ for $i > n$. <u>This is no longer true for a non standard</u> $m$ ; hence we always assume $m$ to be standard in this lesson.

2) A classical formulations of theorem 1 is the following :
Consider the unordered n-tuple of roots of a polynomial $P \in \mathbb{C}_n[x]$ as a point of $\mathbb{C}^n/\mathfrak{S}_n$, where the permutation groupe $\mathfrak{S}_n$ acts by

$$(x_1, \dots, x_n) \longrightarrow (x_{s(1)}, \dots, x_{s(n)}) .$$

Thus we have a mapping $\Phi : \mathbb{C}_n[X] \longrightarrow \mathbb{C}^n/\mathfrak{S}_n$ ; the second set is a metric space for the distance $\inf_{s \in \mathfrak{S}_n} |x_i - x'_{s(i)}|$ and the first is a normed vector space. Theorem 1 asserts that $\Phi$ is continuous and that the unordered $r_i$-tuple of all $(\alpha - a_i)^{r_i}$ is differentiable at $P_o$ (use the non standard characterization of continuity to prove the equivalence).
Theorem 2 may bo formulated along the same lines if $\mathbb{C}$ is replaced by the projective line as to introduce a point at infinity whose halo contains the $m-n$ non finite roots.

3) Theorem 1 has an important consequence :

If all roots of $P_0$ are simple, then the roots of any perturbation $P$ (of the same degree) are simple too and depend differentiably on $P$.

A classical proof of this result uses local inversion for the mapping roots $\longrightarrow$ polynomial. This is not an elementary tool, indeed ; such an "evidence" should have a very simple proof. So, let us write a non-standard proof of theorem 1, and 2 which is really simple.

4) Proof of theorem 1. Call $\alpha_i$ $(1 \leq i \leq n)$ the roots of $P$, distinct or not. Then $P = k(x-\alpha_1)\ldots(x-\alpha_n)$ and $P_0 = k_0(x-a_1)^{r_1}\ldots(x-a_q)^{r_q}$. The numbers $k_0, a_1, \ldots, a_q$ are standard and $k_0 \neq 0$. Moreover $k \sim k_0$.

Choose a standard $u \in \mathbb{C}$ with $u-\alpha_i$ non infinitesimal.

Then $|P_0(u)| \sim |P(u)| = |k||u-\alpha_1|\ldots|u-\alpha_n|$ and no factor is $\sim 0$. Hence, as $P_0(u)$ is standard, every $|u-\alpha_i|$ is finite, and so every $\alpha_i$ is finite ; it has a shadow $°\alpha_i$ and for any standard $z$, we get

$$P_0(z) = °(P(z)) = k_0(z-°\alpha_1)\ldots(z-°\alpha_n) .$$

Thus (transfer), the polynomial $P_0$ has exactly the $x-°\alpha_i$ as factors, which proves the first part of theorem 1.

For the second part, write $P_0(\alpha) = P_0(a) + \frac{(\alpha-a)^r}{r!}(P_0^{(r)}(a) + \eta)$, with $\eta \sim 0$ and $P_0^r(a)$ standard non zero.

Then $(\alpha-a)^r = -\frac{r!\ H(\alpha)}{P_0^{(r)}(a)+\eta}$ ; but as $\alpha \sim a$ and $H \sim 0$, we get $H(\alpha) = H(a) + \|H\|\xi$ with $\xi \sim 0$. Use the finiteness of $\frac{r!\ H(a)}{\|H\|}$ (for $H \neq 0$) to infer

$$(\alpha-a)^r = \frac{-r!\ H(a)}{P_0^{(r)}(a)} + \|H\|\varepsilon$$ with $\varepsilon \sim 0$. The expected linear mapping is

$$L(H) = -\frac{r!\ H(a)}{P_0^{(r)}(a)},$$ which only depends on $P_0$ and $a$.

5) Proof of theorem 2. Call $\alpha_1, \ldots, \alpha_q$ the finite roots of $P$, if any, and $\beta_{q+1}, \ldots, \beta_m$ the infinitely large roots.

Then $P(z) = (z-\alpha_1)\ldots(z-\alpha_q)[k(z-\beta_{q+1})\ldots(z-\beta_m)]$ ; take $z$ standard with $z \neq °\alpha_i$ for every $i$. Then

$$\frac{P_o(z)}{(z - {}^\circ\alpha_1)\ldots(z - {}^\circ\alpha_q)} \sim k(z - \beta_{q+1})\ldots(z - \beta_m) = T(z) .$$

But for $z'$ standard, we have $\frac{T(z')}{T(z)} \sim 1$, because $\frac{z' - \beta_j}{z - \beta_j} \sim 1$ and $m$ is standard ; hence $T(z)$ is nearly constant, which proves that, except for $z = {}^\circ\alpha_i$, $P_o(z) = K(z - {}^\circ\alpha_1)\ldots(z - {}^\circ\alpha_q)$, where $K$ is some constant. This ends the proof (by transfer, the equality is true for <u>all</u> $z \neq {}^\circ\alpha_i$, hence for all $z$ .).

6) Theorem 1 concerns a regular perturbation problem, and theorem 2 a singular one ; we get a very typical behaviour, which preludes to the "layer behaviours" in singular perturbations of differential equations : the solution (here the set of roots) is partly approximated by the solution of the reduced equation $P_o = 0$ and partly "jumps to infinity".

7) Theorem 3 is a partial extension of theorem 1 to the k-variable polynomials. We have to prove that every standard point $a$ on $S_o$ is the shadow of some point on $S$ ; also that any finite point on $S$ has its shadow on $S_o$, which is immediate for $P_o({}^\circ x) \sim P({}^\circ x) \sim P(x) = 0$ (use the fact that $P$ has only finite coefficients). Thus consider all complex straight lines through $a$, with equation $x = a + \lambda u$ . Then $Q^u(\lambda) = P(a + \lambda u)$ is a polynomial in $\lambda$ ; moreover, for any standard $u$, we have $Q^u(\lambda) \sim Q^u_o(\lambda)$ on standard values of $\lambda$ ; choose $u$ such that $Q^u_o$ has degree $\geq 1$ . Then $Q^u$ is a perturbation of the standard polynomial $Q^u_o$, with standard degree. Hence by theorem 2, the root $0$ is the shadow of some root $\lambda$ of $Q^u$ ; thus $a = {}^\circ(a + \lambda u)$ and $P(a + \lambda u) = 0$ ; this ends the proof of theorem 3.

8) <u>Exercises</u>.

1) Assume in the proof of theorem 3 that $a$ is a point of order $n$ (i.e. $n$ is the maximal order of $0$ as a root of $Q^u_o$, whenever $u$ takes all standard values). Improve the result in this case.

2) Let $\omega$ be an infinitely large integer. Consider the polynomial

$$P = \frac{1}{2^\omega \cdot \omega}(z - 1)\ldots(z - \omega) + z^2 .$$

Prove that $P$ is a perturbation of $P_0 = Z^2$ and apply theorem 2 ; infer that $1 = 0$ (of course, this is *not* the simplest proof...).

---

Lesson 2

PERTURBATIONS OF LINEAR OPERATORS ON $K^n$ ($K = \mathbb{R}$ , $\mathbb{C}$ or $\mathbb{H}$)

*Problem. Let* $T_0 : K^n \longrightarrow K^n$ *be a linear map.* Compare the geometry of $T_0$ *with* that of $T_0 + H$ *where* $H$ *is a little perturbation.*

*Formulation within IST* : *Let* $n$ *and* $T_0$ *be standard and* ${}^{\circ}T = T_0$ . *Compare* ...

LEMMA. *If* $n$ *is standard, the shadow* ${}^{\circ}V$ *of a linear subspace* $V$ *of* $K^n$ *is a linear subspace of the same dimension as* $V$ .

THEOREM 1. *If* $n$ *and* $T_0$ *are standard, and if* ${}^{\circ}T = T_0$ , *one has* :

i) *for every subspace* $V$ , $T_0({}^{\circ}V) \subset {}^{\circ}(T(V))$ ; *in particular* $\operatorname{Im} T_0 \subset {}^{\circ}(\operatorname{Im} T)$ *and* $\operatorname{rank} T \geq \operatorname{rank} T_0$ ;

ii) *if* $T(V) \subset V$ , *then* $T_0({}^{\circ}V) \subset {}^{\circ}V$ .

iii) ${}^{\circ}(\operatorname{Ker} T^i) \subset \operatorname{Ker} T_0^i$ *for every* $i \leq n$ . *Hence if* $T$ *is nilpotent, so is* $T_0$ .

iv) *All the eigenvalues of* $T$ *are finite and their shadows are eigenvalues of* $T_0$ ; *if* $x$ *is a finite eigenvector of* $T$ *for an eigenvalue* $\lambda$ , *then* ${}^{\circ}x$ *is an eigenvector of* $T_0$ *for* $\lambda_0$ .

THEOREM 2 (*particular cases*).

i) *If* $K = \mathbb{R}$ *and if* $T$ *is symmetric for a standard inner product on* $\mathbb{R}^n$ , $T_0$ *is symmetric too.*

ii) *If* $K = \mathbb{R}$ *and if* $T$ *is normal* (*resp. hermitian, resp. unitary*) *with respect to a standard hermitian product on* $K^n$ , $T_0$ *has the same property.*

iii) *If* $K = \mathbb{C}$ , *every eigenvalue of* $T_0$ *is the shadow of eigenvalues of* $T$ , *with respect to the multiplicities, as in theorem 1, lesson 1.*

iv) *If* $K = \mathbb{C}$ , *let* $\mathbb{C}^n = E_1 \oplus \ldots \oplus E_r$ *be the characteristic decomposition of* $T$ (*i.e. the characteristic polynomial has roots* $\lambda_i$ *of order* $q_i$ *and*

$E_i = \mathrm{Ker}(T-\lambda_i I)^{q_i}$ ) ; summing up the $E_i$ 's with equivalent $\lambda_i$ 's , one has a decomposition $\mathbb{C}^n = F_1 \oplus \ldots \oplus F_s$ and the shadows $°F_i$ give the <u>characteristic decomposition for</u> $P_0$ .

<u>Comments</u>. 0) The statements above are short and natural formulations covering an important part of linear operator's perturbation theory in finite dimension - even in this case, this theory is not trivial (see KATO).

1) The lemma is an essential tool in lesson 2 and 3 ; its classical meaning is that the Grassman manifolds (i.e. the set of q-planes in $\mathbb{R}^n$ with an appropriate topology) are compact.(Recall lesson III.3 about compactness) ; but we never use this classical equivalent and therefore we don't worry about the topology of the Grassmann manifolds.

<u>Proof of the lemma</u>. Recall that $°V$ is the set in $K^n$ whose standard elements are the shadow of the finite vectors of $V_i$ ; by transfer, we prove any internal property of $°V$ on the standard elements. For instance, if $a$ , $b$ are standard in $°V$ , then $a = °x$ , $b = °y$ with $x, y \in V$ . But addition in $K^n$ is standard (for $n$ is) and continuous ; hence $a+b = °x + °y = °(x+y)$ is in $°V$ . In the same way, we prove that $°V$ is a linear subspace. As for dimension, let $e_1, \ldots, e_r$ be a standard orthonormal basis of $V$ , for some standard inner product on $K^n$ . Use orthonormalization (for $K = \mathbb{C}$ or $\mathbb{H}$ , take $\Sigma \frac{1}{2} (q\bar{q}' + q'\bar{q})$ for instance) ; as $< e_1, e_i > = 1$ , the $e_i$ 's are finite and (continuity of $< , >$ ) the $°e_i$ are orthonormal and generate $°V$ (a finite vector of $V$ has finite components on the $e_i$ 's). Hence $\dim V = \dim V_0$ .

2) " $°T = T_0$ " means that for every standard $x \in K^n$ , $T(x) \sim T_0(x)$ . Thus for a standard basis $\{e_i\}$ , $T(e_i) \sim T_0(e_i)$ and the matrix of $T$ is the shadow of $T_0$ ones. The converse is obvious, for a standard $x$ has standard components. We conclude that for every finite $x$ , $T(x) \sim T_0( x) \sim T_0(°x)$ .

3) The proofs of (i) , (ii) , (iii) in theorem 1 are immediate ; just remark that a nilpotent $T$ is such that $T^i = 0$ for some $i \leq n$ , for if $T^{i-1} \neq 0$

and $T^i = 0$ , the sequence $\mathrm{Ker}\, T \subset \mathrm{Ker}\, T^2 \subset \ldots \subset \mathrm{Ker}\, T^i = K^n$ has strictly growing dimension.

As for (iv), consider $x \in K^n$ and $\lambda \in K$ such that $T(x) = \lambda x$ , $x \neq 0$ . Then $T(\frac{x}{\|x\|}) = \lambda \frac{x}{\|x\|}$ and $\|T(\frac{x}{\|x\|})\| = |\lambda|$ (for a standard norm) ; thus $|\lambda| \sim \|T_o(°(\frac{x}{\|x\|}))\|$ and $\lambda$ is finite ; now $T_o(\frac{x}{\|x\|}) = °(T(\frac{x}{\|x\|})) = °(\lambda \frac{x}{\|x\|})$ $\sim °\lambda\, °(\frac{x}{\|x\|})$ and $°\lambda$ is an eigenvalue of $T_o$ . Furthermore if $T(x) = \lambda x$ with $x$ finite, we have $T_o(°x) = °(T_o(x)) = °\lambda\, °x$ and $°x$ is an eigenvector of $T_o$. Note that we don't know if every eigenvalue of $T_o$ is the shadow of some eigenvalue of $T$ . In the complex case, it's true, for the characteristic polynomials have $n$ roots, and th. 1 of lesson 1 applies.

4) <u>Remark</u>. The proof above is very easy and close to euristic ideas ; you could be tempted to get a classical proof which sounds alike. But as soon as the statement is formulated, the work seems no longer easy ! For instance, try it for the classical equivalent of theorem 1, (IV), that is (for the first part) : "for every $\varepsilon > 0$ , there is an $\eta > 0$ such that for $\|T - T_o\| < \eta$ , every eigenvalue of $T$ is at distance less than $\varepsilon$ from some eigenvalue of $T_o$."
This remark works all along chapter IV ; it's perhaps the most insidious misunderstanding that watches a superficial observer, as long as he is intimely convinced that what is easy with non standard analysis certainly is as easy without it.

5) <u>Proof of theorem 2.</u>

(i) and (ii) are immediate computations on standard elements (use continuity of the standard inner product).

(iii) is an application of th. 1, lesson 1, to the charasteristic polynomials, whose coefficients depend continuously on the matrices.

As for (iv), which is a result on continuity of charasteristic spaces, call $\mu_1, \ldots, \mu_t$ the distinct eigenvalues of $T$ with same shadow $\alpha$ . Then, if $F$ is the direct sum of the $E_i$ 's associated with the $\mu_i$ 's , $°F$ is clearly contained in $\mathrm{Ker}(T_o - \alpha I)^m$ , where $m$ is the order of $\alpha$ . Thus we get subspaces $°F_j$ , the sum of which is direct. But from the lemma above, the dimension of each $°F$ is the sum of the dimensions of the corresponding $E_i$ 's ; Compare dimensions and you get $°F = \mathrm{Ker}(T_o - \alpha I)^m$ for each $j$ .

6) Let us give some funny applications of these results. Consider a standard complex linear operator $T_o$ in $\mathbb{C}^n$ . It's easy to find a standard basis in which its matrix is triangular . take an eigenvector and a supplementary subspace, and so on $n$ times.

Now $T_o = \bar{S}^1 D_o S$ with $S$ standard and $D_o$ triangular. Add some infinitesimals on the diagonal of $D_o$ and you get a matrix $D$ with $n$ distinct eigenvalues and $D \sim D_o$ , which implies $T = \bar{S}^1 D S \sim T_o$ . Note that if $T_o$ is normal with respect to a standard hermitian product, we can get $T$ normal and $S$ unitary (choose orthogonals in the previous procedure). We can use these trivial remarks to put $T_o$ in reduced form. The recipe is the following : choose a $T \sim T_o$ with $n$ distinct eigenvalues $\lambda_i$ as above ; compute an eigenvector basis $\{e_i\}$ for $T$ with unitary vectors (normalize them) ; group the $\lambda_i$ 's of same shadow, and also the corresponding $e_i$ 's ; in such a pack $\{e_1 , \dots , e_t\}$ orthonormalize (for the usual standard product), and you get a basis $\{u_1 , \dots , u_t\}$ of the same subspace with

$$\begin{cases} u_1 = e_1 \\ u_2 = k_2^1 e_1 + k_2^2 e_2 \\ \vdots \\ u_t = k_t^1 e_1 + \dots + k_t^t e_t \quad ; \end{cases}$$

now you have a basis of $\mathbb{C}^n$ in which the matrix $A$ of $T$ is built up of triangular blocs along the diagonal. This basis has a shadow which is a standard basis (easy proof) and in this basis, the matrix $A_o$ of $T_o$ is the shadow of $A$ ; this $A_o$ is then in Jordan form, for the diagonal of each bloc of $A$ contains the eigenvalues of same shadow.

In the normal case, you have the $e_i$ 's orthonormal, $A$ is diagonal, and you get immediately a shadow basis in which $T_o$ is diagonal !

Of course, if you want the fine Jordan form, be a little more careful ..

Remark. This trick can be compared with an analogous technic about standard continuous vector fields which are the shadows of $C^1$-vectors fields (cf. Lesson III.6). In general, for objects which have nice perturbation, a lot of properties are the "shadows" of easy-to-prove properties of the perturbation. Apply this to the well-known genericity theorems and you certainly get with few work some properties of degenerate objects.

REFERENCE.

T. KATO. Perturbation Theory for linear operators. Springer-Verlag.

---

## Lesson 3

## PERTURBATIONS OF LIE ALGEBRA STRUCTURES

DEFINITION. Let $\mu_o$ be a standard Lie algebra structure on $C^n$ ( n standard). A perturbation $\mu$ of $\mu_o$ is a Lie algebra structure on $C^n$ such that for every standard $x, y$ in $C^n$, $\mu_o(x, y) \sim \mu(x, y)$.

PROPOSITION. Let $\mu_o$ be semi-simple (resp. simple). Then every perturbation $\mu$ of $\mu_o$ is semi-simple (resp. simple) and $\mu$ has a Weyl basis whose shadow exists and is a Weyl basis of $\mu_o$ with the same structure constants.

COROLLARY (Nijenhuis-Richardson). Any semi-simple Lie algebra structure on $C^n$ is rigid.

Comments. 0) A Lie algebra structure on $C^n$ is a squewsymmetric bilinear mapping $\mu_o : C^n \times C^n \longrightarrow C^n$ satisfying Jacobi's identity

$$\mu_o(x, \mu_o(y, z)) + \mu_o(y, \mu_o(z, x)) + \mu_o(z, \mu_o(x, y)) = 0 .$$

Let $(x_1, \dots, x_n)$ be a basis of $C_n$ ; the numbers $C^i_{jk}$ such that

$\mu_o(x_j, x_k) = \sum_{i=1}^{n} C_{jk}^i x_i$ are the structure constants of $\mu$. If $n, \mu_o$ and the basis $(x_i)$ are standard, then the $C_{jk}^i$ are also standard and the constants of a perturbation $\mu$ in the same basis have these $C_{jk}^i$ as shadows. Moreover the definition means that $\mu_o$ is the shadow of $\mu$.

Notice that if $\mu_o$ is only supposed to be bilinear and $\mu$ is a perturbation which is squewsymmetric and satisfies Jacobi's identity, so does also $\mu_o$. This is a consequence of the following fact : if $x$ and $y$ are finite and if a bilinear mapping $\mu$ has a shadow (i.e. takes finite values on standard arguments), then $\mu(x, y) \sim \mu(°x, °y) \sim (°\mu)(°x, °y) \sim (°\mu)(x, y)$. (This is an example of S-continuity, as in lesson III.4 ; use a basis to prove it.)

1) There is no classical notion of perturbation. However, statements like "every perturbation of $\mu_o$ has property P ", where P is an internal property, translate into statements like "there is a neighbourhood of $\mu_o$ in which each $\mu$ has property P " ; the topology involved is the natural one on the set of all $\frac{n(n-1)}{2}$-uples of constants $C_{jk}^i$ satisfying $C_{jk}^i = -C_{kj}^i$ and the Jacobi condition. You get an algebraic manifold in $C^{\frac{n(n-1)}{2}}$ endowed with the induced topology. The study of this manifold is an important research area, reserved to cohomologically minded people. Here we make some elementary remarks using simple shadow tricks instead cohomology.

2) Fundamental lemma. Let $V$ be a subalgebra of $C^n$ with respect to $\mu$ ; then $°V$ is a subalgebra with respect to $\mu_o$ with $\dim V = \dim °V$. Moreover, if $V$ is an ideal, so is $°V$.

Proof. From lesson 2, we know that $°V$ is a linear subspace with the same dimension as $V$. Now, take $x_o, y_o$ standard in $°V$. There are elements $x, y$ in $V$ with $x_o = °x$ and $y_o = °y$. Then $\mu_o(x_o, y_o) \sim \mu(x_o, y_o) \sim \mu(x, y)$. As $\mu(x, y)$ $\mu(x, y) \in V$, we get $\mu_o(x_o, y_o) \in °V$. By transfer, $°V$ is stable under $\mu_o$. If $V$ is an ideal, take $x_o$ in $°V$ and $y_o$ in $C^n$, both standard. The same computation yields $\mu_o(x_o, y_o) \in °V$. Hence $°V$ is an ideal. Note that an abelian ideal has an abelian shadow.

Applications. i) If $\mu$ is solvable, so is $\mu_o$ ;

ii) If $\mu$ is nilpotent, so is $\mu_o$ ;

iii) If $\mu_o$ is semi-simple (resp. simple), so is $\mu$ ;

iv) Let $K$ (resp. $K_o$ ) be the Killing-Cartan Form of $\mu$ (resp. $\mu_o$).

Then $K_o = {}^{\circ}K$ .

Proof. Apply the lemma to the following characterizations (see Bourbaki XXVI, chap. I).

- $\mu$ is solvalbe iff there exists a decreasing sequence of ideals $\mathbb{C}^n \supset I_1 \supset \ldots \supset I_p = \{0\}$ $(p \le n)$ such that $\mu(I_i , I_i) \subset I_{i+1}$ .
- $\mu$ is nilpotent iff there exists a decreasing sequence of ideals with $\mu(\mathbb{C}^n , I_i) \subset I_{i+1}$ , $0 \le i \le p \le n$ .
- $\mu$ is semi-simple (resp. simple) iff every abelian ideal (resp. ideal) is $\{0\}$ or the whole space.

In each case, take the shadows of the involved ideals.

As for (iv), use the definition $K(x , y) = \mathrm{tr}(\mathrm{ad}\, x \circ \mathrm{ad}\, y)$ ; for $x , y$ standard, we have $\mathrm{ad}^{\mu_o} x = {}^{\circ}(\mathrm{ad}^{\mu} x)$ and the result follows.

3) Recall that every semi-simple Lie algebra has particular generators $(U_\alpha , Y_\alpha , Z_\alpha)_{\alpha\in\Delta}$ , where $\Delta$ is the set of roots, with the following properties :

$$\begin{cases} [U_\alpha , Y_\alpha] = \alpha(H_\alpha)\, Z_\alpha \quad \text{where} \quad H_\alpha = -i\, U_\alpha \\ [U_\alpha , Z_\alpha] = -\alpha(H_\alpha)\, Y_\alpha \\ [Y_\alpha , Z_\alpha] = 2U_\alpha \, . \end{cases}$$

$\alpha(H_\alpha)$ and all other structure constants only depends on the weights of the roots (which are bounded by $n$ ).

Moreover $\alpha(H_\alpha) \neq 0$ and the Killing-Cartan form takes the following values on the basis :

$$\begin{cases} K(Y_\alpha, Z_\beta) = 0 \\ K(U_\alpha, Z_\beta) = 0 \\ K(U_\alpha, Y_\beta) = 0 \\ K(U_\alpha, U_\alpha) = -\alpha(H_\alpha) \end{cases} \qquad \begin{cases} K(Y_\alpha, Y_\beta) = 0 \text{ if } \alpha \neq \pm\beta \\ K(Y_\alpha, Y_\alpha) = -2 \\ K(Z_\alpha, Z_\beta) = 0 \text{ if } \alpha \neq \pm\beta \\ K(Z_\alpha, Z_\alpha) = -2 \end{cases}$$

If $\{\alpha_1, \dots, \alpha_r)$ is a basis of the root system, then $K(U_{\alpha_i}, U_{\alpha_j}) = 0$ for $i \neq j$ and the vectors $U_{\alpha_i}$, $i = 1, \dots, r$, $Y_\alpha$ with $\alpha > 0$, $Z_\alpha$ with $\alpha > 0$ are independent and generate $\mathbb{C}^n$. This basis is thus orthogonal with respect to $K$; moreover it generates the real compact form $\mathcal{G}^{\mathbb{R}}$ of $(\mathbb{C}^n, \mu) = \mathcal{G}$ and the Killing-Cartan form $K^{\mathbb{R}}$ of $\mathcal{G}^{\mathbb{R}}$ is the restriction of $K$ to $\mathcal{G}^{\mathbb{R}}$. Notice that $K^{\mathbb{R}}$ is negative definite.

Such a basis is called a Weyl basis and $r$ is the rank of $\mathcal{G}$. Any discussion about semi-simple Lie algebras starts with a Weyl basis (see the book of SAGLE-WALDE "Introduction to Lie groups and Lie algebras." Ac.Press, or JACOBSON "Lie algebras." Interscience, for definitions and existence proof.).

Now consider a standard semi-simple structure $\mu_o$ on $\mathbb{C}^n$ ( $n$ standard) and a perturbation $\mu$. As $\mu$ is also semi-simple, it has a Weyl basis ; moreover the weights are standard (they are bounded by a standard number) hence <u>the structure constants in this basis are standard</u>. It is quite natural to ask whether a Weyl basis of $\mu$ is related to some Weyl basis of $\mu_o$. The answer is the best possible as mentionned in the proposition above ; the proof is elementary as we shall see.

• <u>The shadow</u> $°(\mathcal{G}^{\mathbb{R}})$ <u>is a compact real form of</u> $\mathcal{G}_o = (\mathbb{C}^n, \mu_o)$.

<u>Proof</u>. As $\mathcal{G}^{\mathbb{R}}$ is stable under $\mu$, so is $°(\mathcal{G}^{\mathbb{R}})$ under $\mu_o$. From $\mathbb{C}^n = \mathcal{G}^{\mathbb{R}} \oplus i\,\mathcal{G}^{\mathbb{R}}$ (real decomposition), we get $\mathbb{C}^n = °(\mathcal{G}^{\mathbb{R}}) \oplus i\,(°\mathcal{G}^{\mathbb{R}})$ ; hence $°(\mathcal{G}^{\mathbb{R}})$ is a real form of $\mathcal{G}_o$. The Killing-Cartan form $K_o$ of $\mathcal{G}_o$ restricted to $°(\mathcal{G}^{\mathbb{R}})$ is thus <u>non degenerate</u> (semi-simplicity) ; call it $K_o^{\mathbb{R}}$. As every standard $x \in °(\mathcal{G}^{\mathbb{R}})$ is the shadow of some $y \in \mathcal{G}^{\mathbb{R}}$, we have $K_o^{\mathbb{R}}(x, x) \sim K(y, y) < 0$, hence $K_o^{\mathbb{R}}(x, x) \leq 0$. By transfer, infer $K_o(x, x) \leq 0$ for every $x \in °(\mathcal{G}^{\mathbb{R}})$ ; this proves that $K_o^{\mathbb{R}}$ is <u>negative definite</u>, i.e. compactness of the real form.

• <u>The vectors</u> $U_\alpha$ , $Y_\alpha$ , $Z_\alpha$ <u>are finite and not infinitesimal in norm.</u>

Consider some standard norm $\| \ \|$ on $C^n$ and put $a = \|U_\alpha\|$ , $b = \|Y_\alpha\|$ , $c = \|Z_\alpha\|$ .
The vectors $U'_\alpha = \frac{U_\alpha}{a}$ , $Y'_\alpha = \frac{Y'_\alpha}{b}$ , $Z'_\alpha = \frac{Z_\alpha}{c}$ have non zero shadows in $°(\mathcal{G}^{\mathbb{R}})$ .
Now $\mu_o(°U'_\alpha , °Y'_\alpha) \sim \mu(U'_\alpha , Y'_\alpha) = \frac{c}{ab}\,\alpha(H_\alpha)\, Z'_\alpha$ : As $\alpha(H_\alpha)$ is standard $>0$, we get $\frac{c}{ab}$ finite. A similar argument shows that $\frac{b}{ac}$ and $\frac{a}{bc}$ are also finite.
Suppose $\frac{c}{ab} \sim 0$ ; then $\mu_o(°U'_\alpha , °Y'_\alpha) = 0$ ; but

$$K_o^{\mathbb{R}}(\mu_o(°U'_\alpha , °Y'_\alpha) , °Z'_\alpha) = K_o^{\mathbb{R}}(°U'_\alpha , \mu_o(°Y'_\alpha , °Z'_\alpha)) = 2\, °(\tfrac{a}{bc})\, K_o^{\mathbb{R}}(°U'_\alpha , °U'_\alpha) .$$

As $K_o^{\mathbb{R}}(°U'_\alpha , °U'_\alpha) < 0$ , we get $\frac{a}{bc} \sim 0$ .
Similarly $0 = K_o^{\mathbb{R}}(°Y'_\alpha , \mu_o(°U'_\alpha , °Z'_\alpha)) = °(\frac{b}{ac})\, K_o^{\mathbb{R}}(°Y'_\alpha , °Y'_\alpha)\, \alpha(H_\alpha)$ , hence $\frac{b}{ac} \sim 0$ .
Similar arguments prove that whenever one of the numbers $\frac{c}{ab}$ , $\frac{b}{ac}$ , $\frac{a}{bc}$ is $\sim 0$ , both others are also $\sim 0$ . But then the two-by-two products $\frac{1}{a}$ , $\frac{1}{b}$ , $\frac{1}{c}$ are infinitesimal !
Now $K(U'_\alpha , U'_\alpha) = \frac{1}{\alpha^2} K(U_\alpha , U_\alpha) = -\frac{1}{\alpha^2}\,\alpha(H_\alpha) \sim 0$ for $\alpha(H_\alpha)$ is standard ; this is not possible, for $K(U'_\alpha , U'_\alpha) \sim K_o^{\mathbb{R}}(°U'_\alpha , °U'_\alpha)$ would imply $°U'_\alpha = 0$ ( $K_o^{\mathbb{R}}$ is negative definite).
Thus the numbers $\frac{c}{ab}$ , $\frac{b}{ac}$ , $\frac{a}{bc}$ are finite and not $\sim 0$ ; the same is true for $\frac{1}{a}$ , $\frac{1}{b}$ , $\frac{1}{c}$ . Hence $U_\alpha$ , $Y_\alpha$ , $Z_\alpha$ have non zero shadows in $\mathcal{G}_o^{\mathbb{R}}$ .
Moreover the shadow of the Weyl basis $U_{\alpha_i}$ , $Y_\alpha$ , $Z_\alpha$ $(\alpha > 0)$ are orthogonal for the scalar product $-K_o^{\mathbb{R}}$ ; hence they are independent and as $\dim \mathcal{G}_o^{\mathbb{R}} = \dim \mathcal{G}^{\mathbb{R}}$ , we have a real basis for $\mathcal{G}_o^{\mathbb{R}}$ and also a complex basis for $\mathcal{G}_o$ .
Now, as the structure constants of $\mathcal{G}$ in the Weyl basis are standard, they remain unchanged for the shadows $°U_{\alpha_i}$ , $°Y_\alpha$ , $°Z_\alpha$ . Hence we have a Weyl basis for $\mathcal{G}_o$ , with the same roots. Furthermore
<u>In these basis</u> $\mathcal{G}$ <u>and</u> $\mathcal{G}_o$ <u>have the same constants</u>, <u>thus are isomorphic as Lie algebras</u>.

4) <u>Proof of the corollary</u>. We have to prove that any semi-simple Lie algebra structure $\mu_o$ on $C^n$ has a neighbourhood in which every $\mu$ is isomorphic to $\mu_o$ (rigidity of $\mu_o$ ).
By transfer, this is equivalent to : "for $n$ standard, every perturbation of a

standard semi-simple $\mu_o$ is isomorphic to $\mu_o$ ". This ends the proof, after § 3.

Remark. If you ask some specialist of Lie algebras about this question, he first will tell you that rigidity of semi-simple Lie algebras is an evidence, since they are classified : for every n , there is only a finite number of non isomorphic semi-simple structures on $C^n$ ; hence it is clear, due to continuity, that in each isomorphy class the only possible limit points are in the class itself. Then you wonder about continuity versus finiteness ; naively you think that you could approach $\mu_o$ jumping from one class to the other ! You claim for some details...

At this point, you get the following precisions :

i) Everybody knows that any Lie algebra with $H^2(\mathfrak{G},\mathfrak{G}) = 0$ (second cohomology class with values in $\mathfrak{G}$ ) is rigid.

ii) It is well-known that semi-simple Lie algebras satisfy $H^2(\mathfrak{G},\mathfrak{G}) = 0$, hence are rigid.

This continuity versus cohomology is not an evidence for non-algebraic minded people, of course. If you throw an eye in the litterature, you discover that (i) is an important result of Nijenhuis-Richardson (see Journal of Math Mec Vol 17 n° 1) and that (ii) is a complicated consequence of the existence of Weyl basis, using Whitehead Lemmas or alternatively some computations on spectral sequences...

On the whole, algebraic evidence, isn't it ?

Lesson 4

# DEFORMATIONS OF LIE ALGEBRAS

DEFINITION. Let $\mu_1$ be a standard Lie algebra structure on $\mathbb{C}^n$ ( n standard). A deformation $\mu_0$ of $\mu_1$ is a standard bilinear mapping of $\mathbb{C}^n \times \mathbb{C}^n$ into $\mathbb{C}^n$ such that there is a linear isomorphism $h : \mathbb{C}^n \longrightarrow \mathbb{C}^n$ for which, on any standard vectors $x, y$, one has $\mu_0(x, y) \sim h^{-1}\ \mu_1(h(x), h(y))$.

PROPOSITION. $\mu_0$ is a Lie algebra structure on $\mathbb{C}^n$ and $\mu = h^{-1} \circ \mu_1 \circ (h \times h)$ is a perturbation of $\mu_0$. Call $\mu$ a transition from $\mu_1$ to $\mu_0$.

PROBLEM. Let $\mu_0$ be a standard Lie algebra structure on $\mathbb{C}^n$ ( n standard). Characterize all standard $\mu_1$ of which $\mu_0$ is a deformation.

Comments. 0) Changing $\mathbb{C}^n$ for $\mathbb{R}^n$ is immaterial in the definition.

1) The definition above is a non-standard characterization (for standard $\mu_0$ and $\mu_1$ ) of the following classical concept of deformation (also called "contraction" in some papers) : a deformation of $\mu_1$ is any point in the closure of the orbit of $\mu_1$ under the action of the linear group ; instead of a single transition, you have a sequence $\mu_n = h_n^{-1} \circ \mu_1 \circ (h_n \times h_n)$, whose limit is $\mu_0$.

It is quite incommon to find explicit transitions in papers about deformations, because computing with sequences of Lie algebras is somewhat cumbersome... Clearly, any answer to the problem above transfers into an information about all deformations, standard or not.

2) Proof of the proposition. Skewsymmetry and the Jacobi condition may be verified on standard vectors $x, y$ (for $\mu_0$ is standard).
The first is obviously the shadow of the skewsymmetry of $\mu$. The second needs also the fact that $\mu_0(x, y) \sim \mu(x, y)$ is not only true for $x, y$ standard, but

also for all finite $x, y$ (recall lesson 3.).

3) Remarks. - It is clear that a given $\mu_1$ may have a lot of non isomorphic deformations, and also that $\mu_0$ may be a deformation of a lot of non isomorphic $\mu_1$.

- If $\mu_0$ is isomorphic to $\mu_1$, it is a deformation of $\mu_1$ (and $h$ is standard, by transfer, whenever $\mu_0$ and $\mu_1$ are standard).

- If $\mu_0$ is a deformation of $\mu_1$, then $\mu_1$ needs not be a deformation of $\mu_0$.

- If $\mu_1$ is a deformation of $\mu_2$ and $\mu_0$ a deformation of $\mu_1$, then $\mu_0$ needs not be a deformation of $\mu_2$.

- A transition $\mu$ from $\mu_1$ to $\mu_0$ (both standard) is a perturbation of $\mu_0$. But any perturbation of $\mu_0$ is not isomorphic to some standard structure $\mu_1$, excepted in particular cases. Let us give an example :

First notice that the abelian structure $\mu_o = 0$ is a deformation of any standard $\mu_1$ moreover $\mu_0$ is the shadow of some point in any orbit, standard or not, provided the dimension $n$ is standard.

This means that, given any $\mu'$ (standard or not), we may find a linear isomorphism $h$ such that $h^{-1} \circ \mu' \circ (h \times h)$ is a perturbation of $\mu_0 = 0$.

Indeed, put $\alpha = \sup_{\|x\| = \|y\| = 1} \mu'(x, y)$, where $\| \ \|$ is some standard norm on $\mathbb{C}^n$ ; define $h$ by $h(x) = \varepsilon^2 x$, where $0 < \varepsilon \leq \frac{1}{\alpha}$ and $\varepsilon \sim 0$.

Then, for standard $x$ and $y$, we have

$$\|h^{-1}(\mu^1(h(x), h(y)))\| = \varepsilon^2 \|\mu^1(x, y)\| \leq \varepsilon^2 \alpha \|x\| \|y\| \sim 0 .$$

Now, it is well known that for $n \geq 3$, $\mathbb{C}^n$ has infinitely many distinct orbits of Lie algebra structures. Take $n$ standard ; then for any standard finite family of structures, there is another one which is not isomorphic to one of this family. Apply the idealization principle and you get a $\mu'$ which is not isomorphic to some standard structure. But $\mu'$ is isomorphic to some perturbation $\mu$ of $0$ ; this $\mu$ is the expected example.

4) For practical computation of deformations, we use a standard basis

$(e_1, \dots, e_n)$ of $C^n$ and define linear isomorphisms by its matrix. Then the transition $\mu$ is given by $\mu(e_i, e_j) = \Sigma\, c_{ij}^k\, e_k$ and the deformation $\mu_0$ by $\mu_0(e_i, e_j) = \Sigma\, {}^\circ(c_{ij}^k)\, e_k$.

If we write differentials $d_1, d, d_0$ instead of Lie brackets, we use the dual basis $(\omega_1, \dots, \omega_n)$ ; then $d\omega = h^* d((h^{-1})^* \omega)$ is given by

$$d\omega_k = -\Sigma\, c_{ij}^k\, \omega_i \wedge \omega_j \quad \text{and} \quad d_0\omega_k = -\Sigma\, {}^\circ(c_{ij}^k)\, \omega_i \wedge \omega_j .$$

(Recall that $d\omega(x, y) = -\omega(\mu(x, y))$ .)

5) If $\mu_0$ is rigid, our problem has an immediate answer : $\mu_0$ is a deformation of $\mu_1$ if and only if $\mu_0$ is isomorphic to $\mu_1$. Thus, in some sense, classifications up to deformations are extensions of those up to isomorphisms, but they are only "one sided" and "non transitive".

Notice that rigidity is stronger than the fact that any standard $\mu_1$ of which $\mu_0$ is a deformation is isomorphic to $\mu_0$. At the time being, almost all results that may be found in the litterature about deformations concern in fact rigidity.

Our aim is to give some informations along another line ; $\mu_0$ being given, we are looking for a property (P) such that $\mu_0$ is a deformation of $\mu_1$ if and only if $\mu_1$ has property (P).

In other words, we say that $\mu_0$ is a model for the property (P).

Thus, the abelian structure is a model for the universal property "to be a Lie algebra structure".

The properties we have in mind concern the differential characters of the linear forms and systems on a Lie algebra. Let us recall some definitions :

A linear form $\omega$ on a Lie algebra $(C^{2p+1}, \mu, d)$ is called a <u>contact form</u> if $\omega \wedge d\omega^p \neq 0$. A less restrictive condition, valuable in all dimensions $n \geq 2p+1$ is $\omega \wedge d\omega^p \neq 0$ ; we say that <u>the class of</u> $\omega$ <u>is at least</u> $2p+1$.

A bilinear skewsymmetric form $\eta$ on $C^{2p}$ is called <u>exact symplectic</u>, whenever $\eta^p \neq 0$ and $\eta = d\omega$ for some linear form $\omega$.

In every case, there is a canonical form :

If $\omega$ is a contact form, there is a basis $\omega_1, \omega_2, \dots, \omega_{2p+1}$ of $C^{2p+1}$ with $\omega = \omega_1$ and $d\omega = \omega_2 \wedge \omega_3 + \dots + \omega_{2p} \wedge \omega_{2p+1}$.

If $d\omega$ is symplectic, there is a basis $\omega_1, \omega_2, \ldots, \omega_{2p}$ of $C^{2p}$ with $\omega = \omega_1$ and $d\omega = \omega_1 \wedge \omega_2 + \ldots + \omega_{2p-1} \wedge \omega_{2p}$.

Any contact form has a <u>characteristic vector</u> $e$, which is the unique vector such that $\omega(e) = 1$ and $d\omega(e, y) = 0$ for every $y$. In the canonical form above, $e$ is precisely defined by $\omega(e) = 1$, $\omega_2(e) = \ldots = \omega_{2p+1}(e) = 0$.

Assume that $(C^{2p+1}, \mu, d)$ <u>has a contact form</u> $\omega$, <u>with vector</u> $e$ <u>such that any linear form</u> $\alpha$ <u>with</u> $\alpha(e) = 0$ <u>is closed</u> $(d\alpha = 0)$.

Then we get a basis as above with $d\omega_i = 0$ for $i \geq 2$ ; hence up to an isomorphism, there is an unique algebra with this property.

It is called <u>Heisenberg's algebra</u> $H_{2p+1}$. Notice that in $H_{2p+1}$, any form with $\alpha(e) \neq 0$ is a contact form ; thus $H_{2p+1}$ is the algebra with a contact form which is as much abelian as possible. In dual words, it is characterized by the following property : <u>there is a vector</u> $e \neq 0$ <u>such that any</u> 2p-<u>plane</u> $J$ <u>containing</u> $e$ <u>is an ideal</u>, <u>whereas for any</u> 2p-<u>plane not containing</u> $e$, <u>equation</u> $e = [u, v]$ <u>has a solution for every</u> $u \in J$, $u \neq 0$.

In the dual basis, $H_{2p+1}$ is defined by $[e_2, e_3] = [e_4, e_5] = \ldots = [e_{2p}, e_{2p+1}]$

$$[e_2, e_3] = [e_4, e_5] = \ldots = [e_{2p}, e_{2p+1}] = -e_1 ,$$

all other brackets zero.

6) Now assume only that $(C^{2p+1}, \mu_1, d_1)$ has a contact form. Non isomorphic algebras of this type occur, but fortunately, we shall see that $H_{2p+1}$ <u>is a model for this property</u> (P).

<u>Proof</u>. First transfer into the equivalent statement with all ingredients standard, i.e. P and $(\mu_1, d_1)$. Call $d_o$ the (standard) differential in $H_{2p+1}$.

-Assume that $d_o$ is a deformation of $d_1$, with transition $d$. Then the contact form $\omega$ of $H$ is also contact for $d$, because the contact condition is open as regards the differentials. Thus $d$ has property (P) and also $d_1$ which is isomorphic to $d$.

- Assume that there is a contact form on $C^{2p+1}$ for $d_1$. Then, by transfer, there is also a standard contact form $\omega$, with a standard canonical form

$d_1\omega = \omega_2 \wedge \omega_3 + \ldots + \omega_{2p} \wedge \omega_{2p+1}$. Define $h$ by

$$\begin{cases} h^* \omega_1 = \varepsilon^2 \omega_1 , & \varepsilon \sim 0 \\ h^* \omega_i = \varepsilon \, \omega_i , & i \neq 1 . \end{cases}$$

For $d = h^* \, d(h^{-1})^*$ , we get

$$d\omega_1 = \omega_2 \wedge \omega_3 + \dots + \omega_{2p} \wedge \omega_{2p+1}$$

$$d\omega_k = - \frac{1}{\varepsilon} h^* \, (d_1 \omega_k) = - \varepsilon \, \Sigma \, c^k_{ij} \, \omega_i \wedge \omega_j ,$$

where the $c^k_{ij}$ are the structural constants of $d_1$ . As they are standard, we get $d_o$ as a deformation of $d_1$ .

7) An easy improvement of this argument yields a more general model : The direct product $H_{2p+1} \times A_{n-(2p+1)}$ , $A$ abelian, is a model for the property "to have a linear form of class at least $2p+1$ ".

Thus we have a transitive diagramm for n-dimensional algebras :

$$\dots \longrightarrow H_{2p+1} \times A_{n-(2p+1)} \longrightarrow H_{2p-1} \times A_{n-(2p-1)} \longrightarrow \dots \longrightarrow H_1 \times A_{n-1} \longrightarrow A_n ,$$

where the arrows mean "deformable on" and where a given algebra is deformable on all models from the last for which it contains a linear form of high enough class. This is some sort of "sieve" for Lie algebras..;

Other properties may refine that classification. We list some examples, the proof of which is not always so easy as above.

- The algebra $so_{(3)}$ (i.e. $d\omega_1 = \omega_2 \wedge \omega_3$ , $d\omega_2 = \omega_3 \wedge \omega_1$ , $d\omega_3 = \omega_1 \wedge \omega_2$ ) is a model for the property "any non zero linear form on $\mathbb{C}^3$ is a contact form". Indeed, you have a stronger statement : this property is characteristic of $so(3)$ , up to an isomorphism. Moreover, in higher dimensions, there is no algebra of this type (see M. GOZE, C.R.A.S., t. 283 (1976), p. 499.).

- The algebra $B_2 \times B_2$ , where $B_2$ is the Lie algebra of the affine group of the line (i.e. $d\omega_1 = \omega_1 \wedge \omega_2$ , $d\omega_2 = 0$ ) is a model for the property " $\mathbb{C}^4$ has an exact symplectic form and its maximal nilpotent is 2-dimensional".

Again there is a stronger statement : $B_2 \times B_2$ is rigid.

- The algebra $B_2 \times so(3)$ is a model for the property " $\mathbb{C}^5$ has a contact form

and has no abelian ideal of dimension $> 1$ ":
This also is a case of rigidity, indeed.

- The 5-dimensional algebra with $[e_3, e_4] = e_1$ , $[e_3, e_5] = e_2$ , other brackets zero, is a model for the property " $\mathbb{C}^5$ has a linear system of rank 2 and class 5 ".

8) We may enlarge the concept of model as follows, to allow multiple models .

A family F of Lie algebra structures on $\mathbb{C}^n$ is called a multiple model for a property (P) if and only if for every structure on $\mathbb{C}^n$ , to satisfy P is equivalent to have a deformation in F . Moreover, if two different elements of F cannot be deformed one on the other, F is called irreducible.

To prove that for a standard n , some standard family F is a multiple model for an internal property (P) , we have only to prove that for every standard structure on $\mathbb{C}^n$ , to satisfy (P) is equivalent to have a standard deformation in F . Irreducibility has only to be proved between standard elements of F .

9) Our main example of an irreducible multiple model is a counterpart to the Heisenberg model of contact algebras.

We consider on $(\mathbb{C}^{2p}, d_1)$ the property "to have an exact symplectic form" and we get an irreducible multiple model for it.

To avoid heavy computations, we only discuss the 4-dimensional case (see [G] for the general case).

- Consider a standard structure with differential $d_1$ and a standard symplectic form $d\omega$ on $\mathbb{C}^4$ .

We get a standard basis $\omega = \omega_1, \omega_2, \omega_3, \omega_4$ with $d_1\omega = \omega_1 \wedge \omega_2 + \omega_3 \wedge \omega_4$ .

Define h by

$$\begin{cases} h^*\omega_1 = \varepsilon^2\omega_1 \\ h^*\omega_2 = \omega_2 \\ h^*\omega_3 = \varepsilon\,\omega_3 \\ h^*\omega_4 = \varepsilon\,\omega_4 \end{cases}$$

and you get a transition $d = h^* d_1 (h^{-1})^*$ whose shadow $d_o$ exists and satisfies

$$\begin{cases} d_o\omega_1 = \omega_1 \wedge \omega_2 + \omega_3 \wedge \omega_4 \\ d_o\omega_2 = 0 \\ d_o\omega_3 = a\omega_2 \wedge \omega_3 + b\omega_2 \wedge \omega_4 = \omega_2 \wedge (a\omega_3 + b\omega_4) \\ d_o\omega_4 = c\omega_2 \wedge \omega_3 - (1+a)\omega_2 \wedge \omega_4 = \omega_2 \wedge (c\omega_3 - (1+a)\omega_4) \end{cases}$$

where $a, b, c$ are standard complex numbers.

Replacing $\omega_3, \omega_4$ by $\begin{cases} \omega_3 = \alpha_3^3 \omega_3' + \alpha_3^4 \omega_4' \\ \omega_4 = \alpha_4^3 \omega_3' + \alpha_4^4 \omega_4' \end{cases}$

we get $\begin{cases} d_o\omega_3' = \omega_2 \wedge (a'\omega_3' + b'\omega_4') \\ d_o\omega_4' = \omega_2 \wedge (c'\omega_3' - (1+a')\omega_4') \end{cases}$

with $A' = \alpha^{-1} A \alpha$ for the corresponding matrices.

Assuming $\alpha$ unimodular (i.e. $\alpha_3^3\alpha_4^4 - \alpha_4^3\alpha_3^4 = 1$ ), we have $\omega_3' \wedge \omega_4' = \omega_3 \wedge \omega_4$ . Notice that such a change of basis commutes with $h$ ; hence we may consider $\omega_1, \omega_2, \omega_3', \omega_4'$ as the initial basis before transition (provided $\alpha$ is a standard matrix, of course).

If $A' = \alpha^{-1} A \alpha$ , with $|\alpha| \neq 1$ , we have $A' = (\frac{\alpha}{|\alpha|})^{-1} A \frac{\alpha}{|\alpha|}$ and so get an unimodular change of basis (However, in higher dimensions, $\alpha$ has to be in the symplectic group and to reduce $A$ with this restriction is more complicated).

Thus we reduce $A$ to a Jordan form. As the trace of $A$ is $-1$ (due to the Jacobi condition $d\,d\omega_1 = 0$ ), we get two cases (and the procedure yields a standard $\alpha$ , by transfer) :

$$A' = \begin{pmatrix} a & 0 \\ 0 & -(1+a) \end{pmatrix} \quad \text{and} \quad A' = \begin{pmatrix} -\frac{1}{2} & 1 \\ 0 & -\frac{1}{2} \end{pmatrix} .$$

Call $(\mu_o^a , d_o^a)$ the Lie algebra structure in the first case and $(\mu_o^e , d_o^e)$ the exceptionnal case. Thus, we have proved that <u>the standard family</u> $\{\mu_o^a\}_{a \in \mathbb{C}} \cup \{\mu_o^e\}$ <u>is a multiple model</u> (clearly the symplectic form $d_o\omega_1$ is also symplectic for all perturbations of $d_o$ ).

However this family is not irreducible, <u>because</u> $\mu_o^{-\frac{1}{2}}$ <u>is a deformation of</u> $\mu_o^e$ .

Indeed, define a transition by

$$\begin{cases} k^*\omega_1 = \varepsilon\,\omega_1 \\ k^*\omega_2 = \omega_2 \\ k^*\omega_3 = \omega_3 \\ k^*\omega_4 = \varepsilon\,\omega_4 \end{cases}$$

(from now on, forget the primes in $\omega_3'$ , $\omega_4'$ ). That is :

$$\begin{cases} d\omega_1 = \omega_1\wedge\omega_2 + \omega_3\wedge\omega_4 \\ d\omega_2 = 0 \\ d\omega_3 = -\frac{1}{2}\,\omega_2\wedge\omega_3 + \varepsilon\,\omega_2\wedge\omega_4 \\ d\omega_4 = -\frac{1}{2}\,\omega_2\wedge\omega_4\ . \end{cases}$$

As deformations are not transitive, we cannot immediately infer that the family $(\mu_o^a\ ,\ d_o^a)$ is a multiple model, although it is true.

To prove this, suppose that in the procedure above, some standard $(\mu_1\ ,d_1)$ gives $(\mu_o^e\ ,d_o^e)$ as a deformation. The transition (expressed in the basis of the reduced form) satisfies

$$\begin{cases} d'\omega_1 = \omega_1\wedge\omega_2 + \omega_3\wedge\omega_4 \\ d'\omega_2 = O(\varepsilon) \\ d'\omega_3 = O(\varepsilon) - c_{23}^3\,\omega_2\wedge\omega_3 - c_{24}^3\,\omega_2\wedge\omega_4 \\ d'\omega_4 = O(\varepsilon) - c_{23}^4\,\omega_2\wedge\omega_3 - c_{24}^4\,\omega_2\wedge\omega_4\ , \end{cases}$$

where $O(\varepsilon)$ contains all terms with a factor $\varepsilon$ . As the shadow of $d'$ is $d_o^e$ , we get

$$\begin{cases} c_{23}^3 = \frac{1}{2}\ , & c_{24}^3 = -1 \\ c_{23}^4 = 0\ , & c_{24}^4 = \frac{1}{2}\ . \end{cases}$$

Hence $d'' = k^*\,d'(k^{-1})^*$ satisfies

$$\begin{cases} d''\omega_1 = \omega_1\wedge\omega_2 + \omega_3\wedge\omega_4 \\ d''\omega_2 = O(\varepsilon) \\ d''\omega_3 = O(\varepsilon) - \frac{1}{2}\,\omega_2\wedge\omega_3 + \varepsilon\,\omega_2\wedge\omega_4 \\ d''\omega_4 = O(\varepsilon) - \frac{1}{2}\,\omega_2\wedge\omega_4\ , \end{cases}$$

and $d_o^{-\frac{1}{2}}$ is the shadow of $d''$ as expected. <u>Thus the standard family</u> $(\mu_o^a\ ,d_o^a)$

is a multiple model for the property "To have an exact symplectic form".

10) Now we prove that this family is irreducible.

Indeed, assume $a$ and $a'$ standard such that there is some transition $(\mu, d)$ from $(\mu_o^{a'}, d_o^{a'})$ to $(\mu_o^a, d_o^a)$.

Call $\{e_i\}$ the (standard) dual basis of $(\omega_i)$ and notice that for any $b$, $\mu_o^b$ has an unique abelian ideal $I_b$ of dimension 1, which is generated by $e_1$ and whose normalizer $c = \{x \in C^4, [x, I] = \{0\}\}$ is an ideal generated by $(e_1, e_3, e_4)$.

Call $I$ the corresponding ideal for $\mu$ and $c$ its normalizer. Then $°I$ is a abelian ideal of dimension 1 for $\mu_o(a)$, that is $°I = I_a$. Call $u_1$ a vector of $I$ such that $°u_1 = e_1$. Then $°c \subset c_a(°I)$, thus $c_a(I_a)$ has dimension 3 or 4. But as $\mu_o^a(e_1, e_2) \neq 0$, dimension 4 is not possible. Hence $°c = c_a(°I)$ is generated by $e_1, e_3, e_4$. Call $u_3, u_4$ vectors in $c$ such that $°u_3 = e_3$ and $°u_4 = e_4$, and put $u_2 = e_2$. As $\{x_1, x_2, x_3, x_4\}$ is a standard basis, we infer that $\{u_1, u_2, u_3, u_4\}$ is a basis of $C^4$; it has the following brackets for $\mu$.

$$\begin{cases} \mu(u_1, u_3) = \mu(u_1, u_4) = 0 \\ \mu(u_3, u_4) = \alpha u_1 = \bar{u}_1 \quad \text{with} \quad \alpha \sim 1 \\ \mu(\bar{u}_1, u_2) = \alpha_1 \bar{u}_1 + \varepsilon_1 u_2 + \varepsilon_3 u_3 + \varepsilon_4 u_4 \\ \mu(u_2, u_3) = \varepsilon_4 \bar{u}_1 + \varepsilon_5 u_2 + \alpha_2 u_3 + \varepsilon_6 u_4 \\ \mu(u_2, u_4) = \varepsilon_7 \bar{u}_1 + \varepsilon_8 u_2 + \varepsilon_9 u_3 + \alpha_3 u_4 \end{cases}$$

with $\varepsilon_i \sim 0$ and $\alpha_i \sim 1$.

Writing that $c$ is an ideal, we get $\varepsilon_1 = \varepsilon_5 = \varepsilon_8 = 0$. Now replace $u_2$ by $\bar{u}_2 = \frac{1}{\alpha} u_2$, $u_3$ by $\bar{u}_3 = u_3 + \varepsilon_4 \bar{u}_1$ and $u_4$ by $\bar{u}_4 + \varepsilon_7 \bar{u}_1$. You get a basis which has again the $X_i$'s as shadows. In the dual basis the structural equations of $(\mu, d)$ are :

$$\begin{cases} d\bar{\omega}_1 = \bar{\omega}_1 \wedge \bar{\omega}_2 + \bar{\omega}_3 \wedge \bar{\omega}_4 \\ d\bar{\omega}_2 = 0 \\ d\bar{\omega}_3 = -\bar{\varepsilon}_2 \bar{\omega}_1 \wedge \bar{\omega}_2 - \bar{\alpha}_2 \bar{\omega}_2 \wedge \bar{\omega}_3 - \bar{\varepsilon}_9 \bar{\omega}_2 \wedge \bar{\omega}_4 \\ d\bar{\omega}_4 = -\bar{\varepsilon}_3 \bar{\omega}_1 \wedge \bar{\omega}_2 - \bar{\varepsilon}_6 \bar{\omega}_2 \wedge \bar{\omega}_3 - \bar{\alpha}_3 \bar{\omega}_2 \wedge \bar{\omega}_4 . \end{cases}$$

As $\overline{\omega}_i \sim \omega_i$ and $d \sim d_o^a$ , infer that $\overline{\varepsilon}_i \sim 0$ for $i = 2, 3, 6, 9$ and $\overline{\alpha}_2 \sim -a$ , $\overline{\alpha}_3 \sim 1 + a$ .

Moreover, by $d \circ d = 0$ , we get $\overline{\varepsilon}_2 = \overline{\varepsilon}_3 = 0$ and $1 + \overline{\alpha}_2 + \overline{\alpha}_3 = 0$ .

Thus, the isomorphic structures $\mu$ and $\mu_o^{a'}$ have as eigenvalues for $\mathrm{ad}\, \overline{U}_2|_{\{\overline{U}_3, \overline{U}_4\}}$ and $\mathrm{ad}\, X_2|_{\{X_3, X_4\}}$ respectively the numbers $(a + \varepsilon' , -1 - a + \varepsilon'')$ and $(a' , a'')$ (with $\varepsilon' \sim \varepsilon'' \sim 0$ ). It is easy to check that in all basis where $d\omega_1$ and $d\omega_2$ are reduced as above, these eigenvalues are invariant under isomorphisms ; hence $a' \sim a$ , which implies $a' = a$ (both are standard).

11) In the real case, the reduction of the matrix $\begin{pmatrix} a & b \\ c & -(1+a) \end{pmatrix}$ yields three cases :

$$\begin{pmatrix} a & 0 \\ 0 & -(1+a) \end{pmatrix} , \quad \begin{pmatrix} -\frac{1}{2} & b \\ -b & -\frac{1}{2} \end{pmatrix} \quad \text{and} \quad \begin{pmatrix} -\frac{1}{2} & 1 \\ 0 & -\frac{1}{2} \end{pmatrix} .$$

The latter reduces to the first (for $a = -\frac{1}{2}$) as in the complex case and we get a multiple model with two families. One proves its irreducibility along the same lines as above.

12) Some properties are strong enough to characterize a model $\mu_o$ up to isomorphism. If the property is "open", we get a rigid structure which is clearly a model for it. For instance $so(3)$ , $B_2 \times so(3)$ , $B_2 \times B_2$ are of this type as regards suitable properties.

13) Exercises.

- Prove that "diagonal" transitions with powers of the same $\varepsilon$ as coefficients may always be composed as in the proof of § 9.
- Prove that Heisenberg's algebra is the only model for "contact algebras".

14) Remark. In this lesson, we use N.S.A. as a pleasant tool to deal with deformations, but clearly all explicit transitions may be replaced without problem by sequences. However some proofs would be uneasy, at least in general dimension. Notice that we didn't use the permanence principle in these algebraic lessons, but only very simple properties of infinitesimals, that is a little part of N.S.A.'s power.

Lesson 5

# SLOW-FAST FLOWS IN THE PLANE

<u>Problem. Describe the asymptotic behaviour of the solutions of system</u>

$$(S_\varepsilon) \quad \begin{cases} \varepsilon \dot{x} = y - f(x) \\ \dot{y} = g(x, y) \\ \varepsilon > 0 \;,\; f, g \;\text{ of class } C^1 \end{cases}$$

<u>as</u> $\varepsilon$ <u>tends to</u> $0$ .

<u>Formulation within I.S.T.</u> <u>Assume</u> $f$ <u>and</u> $g$ <u>standard</u>. <u>Describe the eventual shadow of the solutions of</u> $(P_\varepsilon)$ <u>as</u> $\varepsilon \sim 0$ .

<u>Main Lemma. The following symbolic pictures describe in some basic cases the behaviour in a compact standard rectangle.</u>

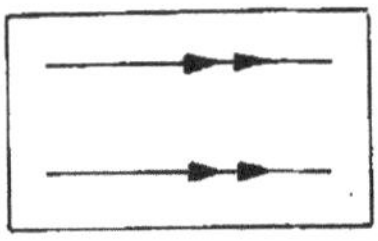

Fig. 1

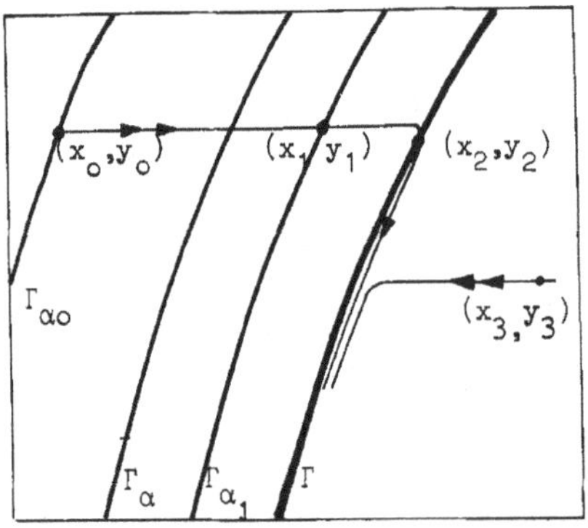


Fig. 2

g<0

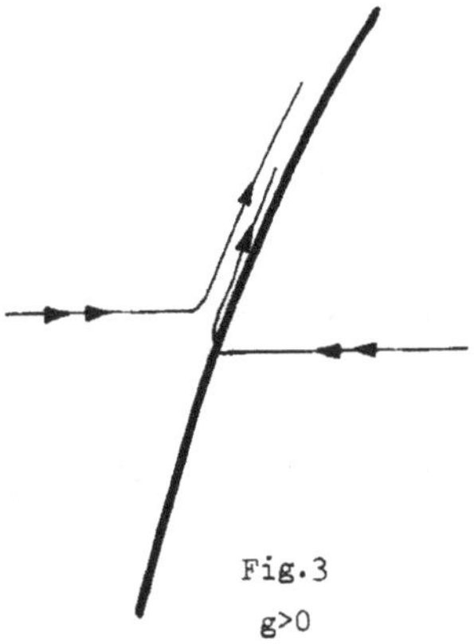

Fig. 3

g>0

Fig. 1.    g<0

Fig.5    g>0

Fig.6    g<0

Fig.7    g>0

Fig.8    g<0

Fig.9    g>0

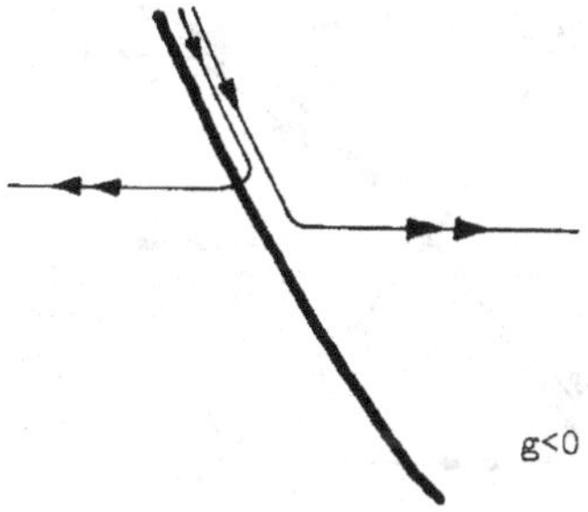

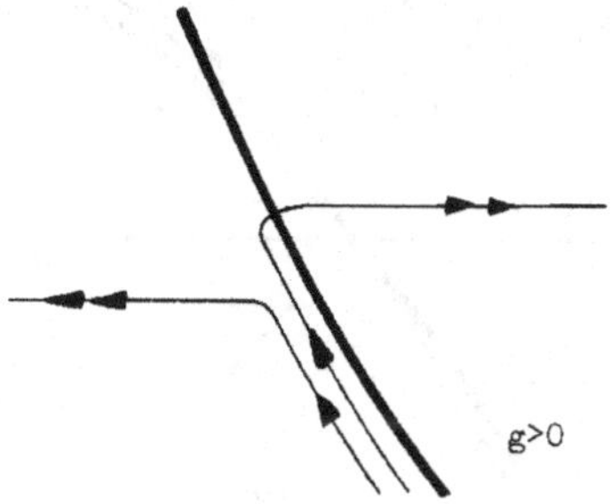

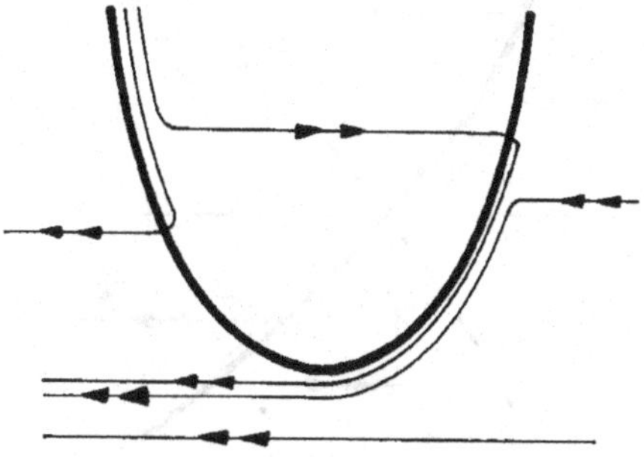

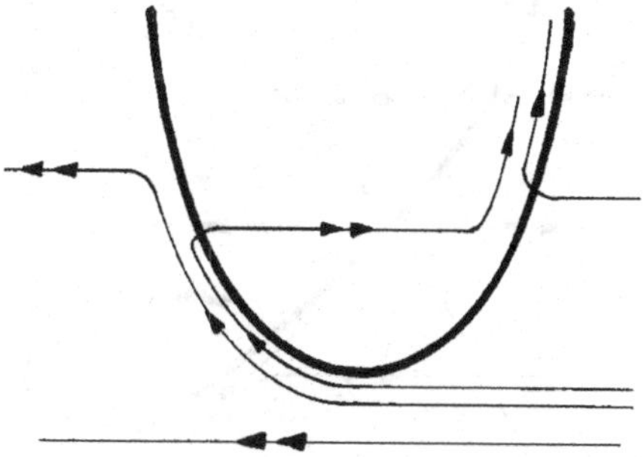

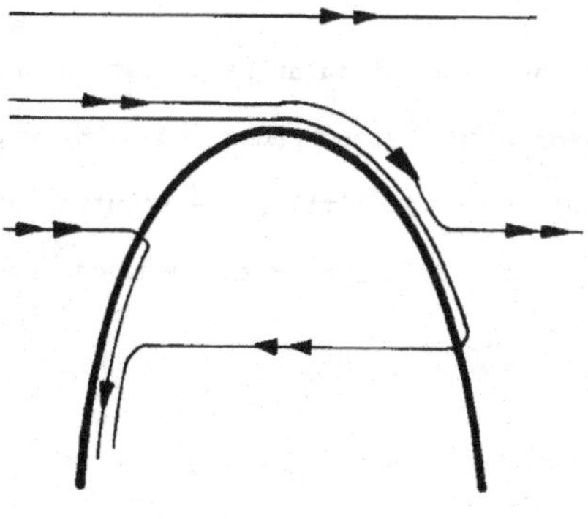

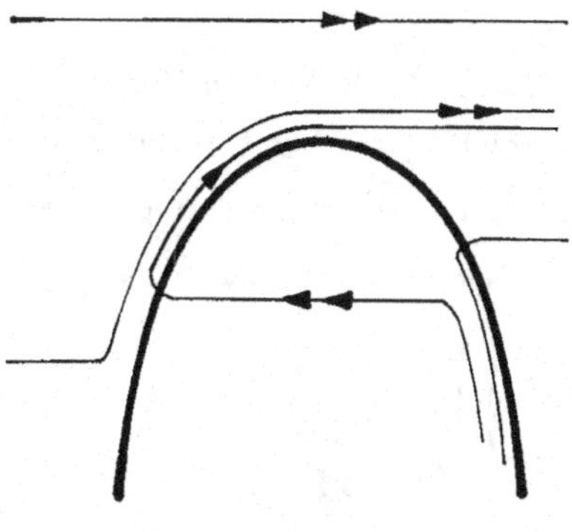

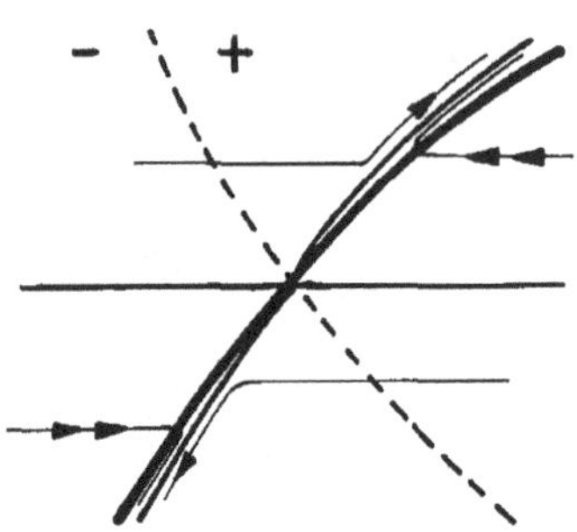


Fig.10

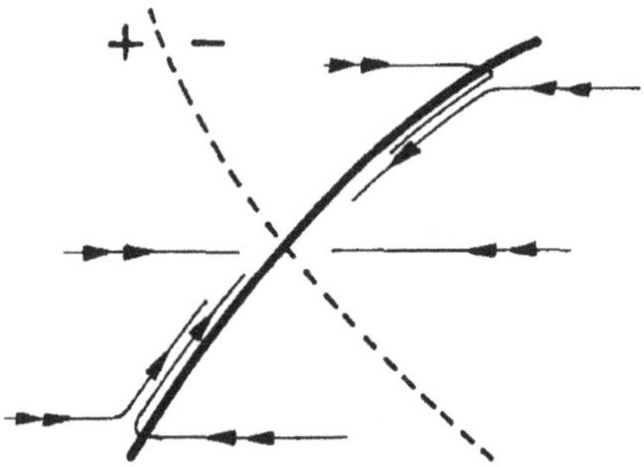


Fig.11

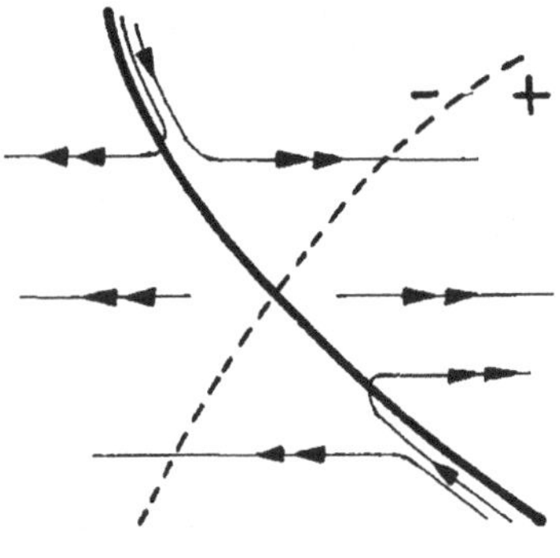


Fig.12

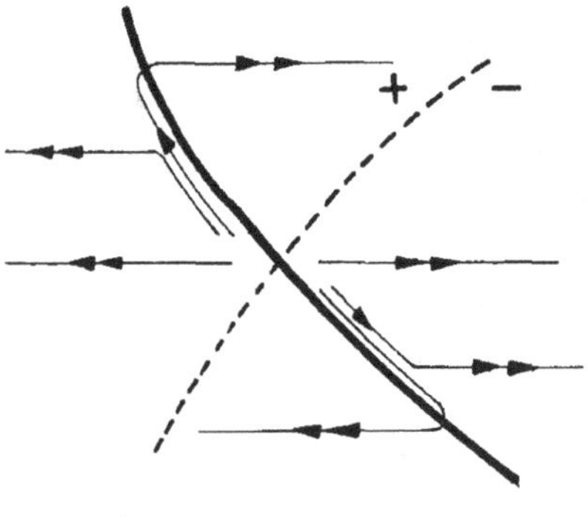


Fig.13

Comments. 0) We discuss in this lesson a very important singular perturbation problem, which is the key to solve further problems about second order equations with a small parameter. What is typical here is that if $\varepsilon$ is little, the solution of $(S_\varepsilon)$ starting at $(x_o, y_o)$ may quickly jump to the solution of the reduced problem

$$\begin{cases} y = f(x) \\ y' = g(x, y) \\ y(0) = y_o \end{cases}$$

and then go along it slowly (i.e. nearly with the same speed as in the reduced problem) until some point where it moves quickly away ; all quick jumps occur nearly on horizontals.

Thus we call the flow associated with $(S_\varepsilon)$ <u>a slow-fast flow</u>. In fact, we should precisely define this concept in the context of a family of flows, as an asymptotic behaviour ; but the formulation would be uneasy : "given a neighbourhood of ..., there is an $\varepsilon_0$ such that for $\varepsilon < \varepsilon_0$ , the solution ..." ; in any case, the flavour of the intuitive description is lost and the prove rather cumbersome, due to the "corner problem" where approximations have to be matched.

In order to use a non standard description, let us precise in this particular situation the discussion about asymptotic behaviours we had in lesson IV.0.

An asymptotic behaviour of a family $(x_\varepsilon , y_\varepsilon)$ of solutions is a statement like

$$\forall f \forall g \forall u_1 \ldots \forall u_n \exists \varepsilon_0 \forall \varepsilon \; ((\varepsilon < \varepsilon_0) \wedge A(u_1,\ldots,u_n , f , g , \varepsilon) \Longrightarrow B(u_1,\ldots,u_n , x_\varepsilon , y_\varepsilon)) ,$$

where $u_1 , \ldots , u_n$ are auxiliary "moving" data, like neighbourhoods, etc... related with $(S_\varepsilon)$ by property $A$ , and where $B$ is the conclusion about the solution.

Within I.S.T., this statement is equivalent to its transfered form

$\forall^{St} f \, \forall^{St} g \, (\ldots)$ . As $A$ and $B$ are internal formulas, by a second transfer, " $\exists \varepsilon_0 \forall \varepsilon ((\varepsilon < \varepsilon_0) \wedge A \Longrightarrow B)$ " is equivalent to " $\exists^{St} \varepsilon_0 \forall \varepsilon ((\varepsilon < \varepsilon_0) \wedge A \Longrightarrow B)$ " (with $f$ and $g$ standard) . Hence for $\varepsilon \sim 0$ , we have $A(\varepsilon) \Longrightarrow B$ ; conversely, if $A(\varepsilon) \Longrightarrow B$ is true for every $\varepsilon \sim 0$ , " $\exists \varepsilon_0 \forall \varepsilon ((\varepsilon < \varepsilon_0) \wedge A \Longrightarrow B)$ " is true (take $\varepsilon_0 \sim 0$ ) ; by transfer, we get a standard $\varepsilon_0$ .

Thus in its transfered form (all data like $f , g$ standard), <u>any asymptotic behaviour of</u> $(x_\varepsilon , y_\varepsilon)$ <u>may be observed and proved restricting</u> $\varepsilon$ <u>to infinitesimal values</u>.

In engineers' words : what is true for $\varepsilon \sim 0$ is true for little $\varepsilon$ .

Now, for $\varepsilon \sim 0$ , a further use of the transfer principle may translate " $A \Longrightarrow B$ " into some "shadow-statement" very close to the intuitive formulation of the behaviour.

For instance, the description above is a precise one, provided the data are standard, $\varepsilon \sim 0$ , and "nearly","quickly" are replaced by "infinitely close to", "with infinitely large speed", etc... Thus in this case, we know precisely what is <u>a</u> slow-

fast flow, without the need of a family of flows.
Note that the solution of $(S_\varepsilon)$ with "moving" initial point $(x_o(\varepsilon), y_o(\varepsilon))$ corresponds in the non standard formulation (with $f, g$ standard) to a non-standard initial point (finite or not). Finiteness means that $(x_o(\varepsilon), y_o(\varepsilon))$ is bounded as $\varepsilon$ tends to $0$. In case the initial point is not finite, some stretching change of scale (a "telescope") may help to see what is going on, using again the main lemma.

Remark. Although our descriptions are easy to translate in classical words, you must have some training to avoid any mistake ; in lesson IV.0. we gave the tools, but training should be got from section III, of course.

1) The main lemma describes the behaviour of $(x_\varepsilon, y_\varepsilon)$ for a fixed $\varepsilon \sim 0$, $f, g$ standard and $x_o, y_o$ finite. Our photographer chosed some views in a standard compact rectangle $K$, which show different non degenerate cases.
The symbolism is the following : the curve $\Gamma$ of equation $y = f(x)$ is drawn as a thick line ; in fig. 10, 11, 12, 13, the dotted line is the curve of equation $g(x, y) = 0$. Other lines are solutions of $S_\varepsilon$ ; the symbol $\longrightarrow\!\!\!\!\!\!\rightarrow\!\!\!\!\rightarrow$ means "horizontal geometric shadow , with infinitely large speed along it" ; a symbol like  shows a solution moving in the halo of $\Gamma$ with its vertical speed infinitely close to $g(x, f(x))$. In the halo of a corner, however, the symbols $\longrightarrow$ or $\twoheadrightarrow$ no longer apply, for the speed makes there the transition between infinitely large and finite values. Also notice that nearly horizontal curves along a tangent to $\Gamma$ (fig. 6 to 9) may have finite speed at some points in the halo of the maximum or minimum of $f(x)$.

Our proofs only use the classical properties of flows : uniqueness of integral curves starting at a given point, continuous dependance on the vector field (i.e. infinitely close fields have infinitely close integral curves al long as the time is finite ; see lesson III.7).
As a typical non-standard tool, we need the permanence principle in its special form (also called Robinson's lemma), which makes the transition at corners clear.
We detail carefully the proof in case 1 and 2 and then only point out what is new

in other cases, the arguments being similar.

Call $Z$ the vector field associated with $(S_\varepsilon)$ and $\gamma(t) = (x(t), y(t))$ an integral curve starting at $(x_o, y_o)$.

2) Case 1. $y > f(x)$ in $K$ ; no assumption on $g$. As $y - f(x)$ has a non infinitesimal minimum on $K$ (use compactness and standardness of $K$ and $f$), and as $|g(x,y)|$ has a finite upper bound, the field $Z$ is nearly horizontal with $\|Z\|$ infinitely large of order $\frac{1}{\varepsilon}$ ; hence $y(t) \sim x_o$ and the time spent by $\gamma(t)$ in $K$ is infinitesimal (of order $\varepsilon$).

Case 2. $g(x,y) < 0$, $f(x)$ strictly increasing and $\Gamma$ crossing $K$. In this case $Z$ is downwards, vertical on $\Gamma$, from left to right over $\Gamma$, from right to left under $\Gamma$ (i.e. $\Gamma$ is an attractor), and nearly horizontal outside of the halo of $\Gamma$.

By compactness of $K$ and continuity of $g$ and $\frac{df}{dx}$, both take only finite and non infinitesimal values on $K$.

Call $\Gamma_\alpha$ the translated curve of equation $y = f(x-\alpha)$. Now the proof :

- start with $y_o - f(x_o) > 0$, not $\sim 0$. There is an unique $\alpha_o < 0$, not $\sim 0$, such that $y_o = f(x_o - \alpha_o)$ ; for each standard $\alpha$ with $\alpha_o < \alpha < 0$, $\gamma$ is of type 1 on the left of $\Gamma_\alpha$. Precisely $\gamma$ meets $\Gamma_\alpha$ after a time $t_\alpha \sim 0$ and for $0 \le t \le t_\alpha$, $y(t) \sim y_o$, $y(t) > f(x(t))$.

This external property is permanent until some negative $\alpha_1 \sim 0$ and at time $t_{\alpha_1} \sim 0$, $\gamma(t) = (x_1, y_1)$ with $y_1 > f(x_1)$, $y_1 \sim f(x_1)$, $x_1 \sim x_o$, $y_1 \sim y_o$. This point is of the next type.

- start at $(x_1, y_1)$ with $y_1 > f(x_1)$ and $y_1 \sim f(x_1)$.

Let $t$ be standard $> 0$ ; then $y(t) - y_1$ is $< 0$, not $\sim 0$, for $g$ is everywhere not $\sim 0$. Suppose that for any $s \le t$, $y(s) - f(x(s)) > 0$ ; then x

$$x_1 - x(t) > x_1 - f^{-1}(y(t)) \sim f^{-1}(y_1) - f^{-1}(y(t)) .$$

As $\frac{df}{dx}$ is not $\sim 0$, we get $x_1 - x(t) > 0$, not $\sim 0$; hence at some $0 < s < t$, $x'(s)$ has to be $< 0$, i.e. $y(s) - f(x(s)) < 0$, which is a contradiction. Thus $\gamma$ meets $\Gamma$ between the times $0$ and $t$. This internal property is permanent until

some $t_0 \sim 0$ ; let $(x_2, y_2)$ be the corresponding intersection point. As $g$ is finite, we have $y_2 - y_1 \leq t_0 x$ finite $\sim 0$ and hence after an infinitesimal time, the curve $\gamma$ starting at $(x_0, y_0)$ reaches $\Gamma$ at $(x_2, y_2)$ with $y_2 \sim y_0$.
As $Z$ is vertical downwards on $\Gamma$, the intersection point is unique and the curve never meets again $\Gamma$.

- Start at $(x_3, y_3)$ with $y_3 - f(x_3) < 0$, not $\sim 0$.
Using the same argument as above, we see that after an infinitesimal time, $\gamma(t)$ is at $(x_4, y_4)$ with $y_4 \sim f(x_4)$, $y_4 < f(x_4)$ and $y_4 \sim y_3$.
Now, in both cases, we have to start in the halo of $\Gamma$.

- Start at $(x_5, y_5)$ with $y_5 \sim f(x_5)$, $y_5 < f(x_5)$. For any standard $\alpha > 0$, the curve $\gamma(t)$ moves strictly between $\Gamma$ and $\Gamma_\alpha$ (on $\Gamma_\alpha$, $Z$ is nearly horizontal).
This internal property is permanent until some $\alpha \sim 0$. Thus $\gamma(t)$ never leaves the halo of $\Gamma$, within $K$, i.e. $y(t) \sim f(x(t))$. Moreover

$$\dot{y}(t) = g(x(t), y(t)) \sim g(x(t), f(x(t))) ,$$

which proves that the vertical speed is slow and nearly the same as in the reduced system.

Cases 3, 4, 5 are variants of case 2. In 4, 5, the description of the integral curves is obtained from the end point, using the arguments of case 2 for negative times.

Cases 6, 7, 8, 9. Here $f$ has a minimum (or maximum) in $K$. In 6, the curves starting under $\Gamma$ cannot go upwards, hence they must leave $\Gamma$ in the halo of the minimum (whenever they reached the halo of $\Gamma$). In 8 and 9, again come back from the end point to describe the behaviour.
In these cases, we have an attractive and a repulsive arc in $\Gamma$ ; but notice that no integral curve goes slowly along both. In Lesson 8 we shall describe a situation where this happens.

Cases 10, 11, 12, 13. Here $\Gamma$ meets transversaly the curve $\Delta$ of equation $g(x, y) = 0$ and $\Gamma$ is strictly monotoneous in $K$. We have a non degenerate singular point of various types ; in cases 10, 12, 13, there is a separatrix. The

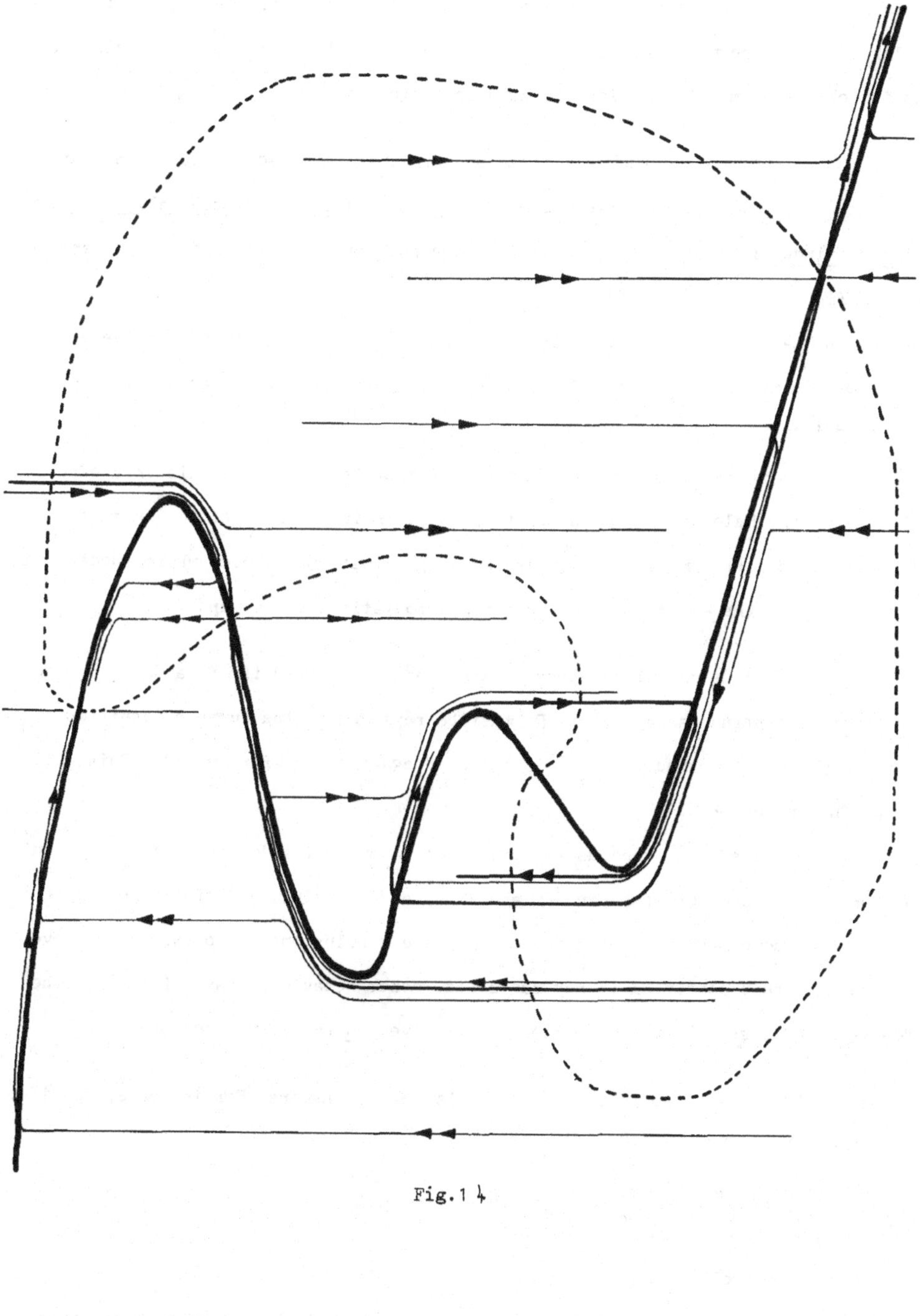

Fig.1 4

description is simply a consequence of 2 , 3 , 4 , 5 outside of the halo of the singular point. By permanence, we get the junction with the behaviour near this point, as it follows from the classical local description.

3) Putting together all these pictures, we may describe the behaviour of the integral curves in any case where $\Gamma$ and $\Delta$ intersect transversaly but not at some singular point of $f$ . This is a very pleasant exercise and we give an example in fig. 14.

Notice that - we need only rough informations on $f$ and $g$ , like relative positions of minima or maxima for $f$ , and relative positions of $\Delta$ and $\Gamma$ , to get the description.

- to match behaviours in two different "photographies" is very easy : what is in the halo of $\Gamma$ in one picture is again in the halo of $\Gamma$ in another. This is quite different from the classical approximation technics where successive steps increase the error and need precise computations on $f$ and $g$ .

4) The descriptions above remain nearly unchanged if $f$ and $g$ depend on $\varepsilon$ in a continuous way ; but $\Gamma$ must be replaced by the curve of equation $y = f(x , y , \varepsilon)$ . The shadow of $\Gamma$ is again of equation $y = f(x , y , 0)$ . This will be useful in lesson 6.

Also in some cases, $g$ depends on some other parameter, independant from $\varepsilon$ . But as long as $f$ and $g$ are standard and $C^1$ , and the singular points are not in the halo of some maximum or minimum of $\Gamma$ , the conclusions are obtained as above. However, if this condition is not satisfied, other behaviours may arise, like the "canards" of lesson 8, which go along attractive and repulsive arcs of $\Gamma$ .

5) We may also be interested in higher dimensions. For instance, consider a system like

$$\begin{cases} x' = \frac{1}{\varepsilon}(z - f(x , y)) \\ y' = g(x , y , z) \\ z' = h(x , y , z) , \end{cases}$$

with $f , g , h$ standard. Such systems occur in the study of non autonomous second order differential equations (for instance in lesson 11 and 14).

Here the curve $\Gamma$ is replaced by the surface of equation $z = f(x, y)$ and provided the singular points are not degenerate, we get again fast nearly horizontal integral curves attracted (or repulsed) by the surface followed (or preceeded) by slow arcs moving in its halo. In lesson 10, we prove that the shadow of these slow arcs is a solution of the reduced system

$$\begin{cases} z = f(x, y) \\ y' = g(x, y, z) \\ z' = h(x, y, z) \end{cases}$$

provided $f, g, h$ are regular enough.

6) A last improvement is to replace $y - f(x)$ by $f(x, y)$ ; then a special study is necessary at points of $\Gamma$ with vertical or horizontal tangent. Some auxiliary change of variables may be useful.

Exercise. Try to describe the integral curves

$$\begin{cases} x' = \frac{1}{\varepsilon}(x^2 + y^2 - 1) \\ y' = ax + by + c \ , \end{cases}$$

where $a, b, c$ are auxiliary parameters.

# Lesson 6

## BOUNDEDNESS OF INTEGRAL CURVES IN EQUATION $\ddot{x} + f'(x)\dot{x} + x = 0$

THEOREM. Let $f$ be an odd function, of order $n$ at $\infty$ (i.e. $\lim_{x \to \infty} f(x)/x^n = k$ with $k \neq 0$ and $n > 1$) ; then every integral curve of equation

$$\text{(I)} \quad \ddot{x} + f'(x)\dot{x} + x = 0$$

is positively bounded.

Comments. 0) Equation (I) describes the motion along a straight line of a material point of mass one under a pull-back central force $-x$ whose action is slacken or quickened by the line's viscosity, supposed to be symmetric around the center.

Wether or not every integral curve is bounded as $t \to \infty$ is an essential property of such mechanical systems, which have been extensively studied, sometimes with much effort ; for $f'(x) = x^2 - 1$, we get ordinary Van der Pol's equation (in opposition to the "singular" one of next lesson).

Our aim is to deduce boundedness as an immediate consequence.

1) The main trick is to transform equation (I) into a system with high speed flow. First use the classical Lienard's transform to get system (II)

$$\begin{cases} \dot{x} = y - f(x) \\ \dot{y} = -x \\ \dot{t} = 1 . \end{cases}$$

With the new variables $\begin{cases} X = \varepsilon x \\ Y = \varepsilon^n y \\ T = \varepsilon^{n-1} t \end{cases}$ , we get system

$$\text{(III)}\quad \begin{cases} \dot{X} = \dfrac{1}{\varepsilon^{2(n-1)}}\left(Y - \varepsilon^n f\left(\dfrac{X}{\varepsilon}\right)\right) \\ \dot{Y} = -X \\ \dot{T} = 1 \, . \end{cases}$$

From $f(x) = kx^n + x^n g(x)$ with $\lim\limits_{x \to \infty} g(x) = 0$, we put it in final form

$$\begin{cases} \dot{X} = \dfrac{1}{\varepsilon^{2(n-1)}}\left(Y - kX^n - g\left(\dfrac{X}{\varepsilon}\right) X^n\right) \\ \dot{Y} = -X \\ \dot{T} = 1 \, . \end{cases}$$

2) Proof of theorem. i) Transfer, that is suppose $f, n, k$ standard ;

ii) Take $\varepsilon \sim 0$ and you get a high speed flow system ; call $K$ the standard unit square around the origin in the $(X, Y)$-plane and you are in case $(k>0)$ or $(k<0)$ of the main lemma, lesson 5.

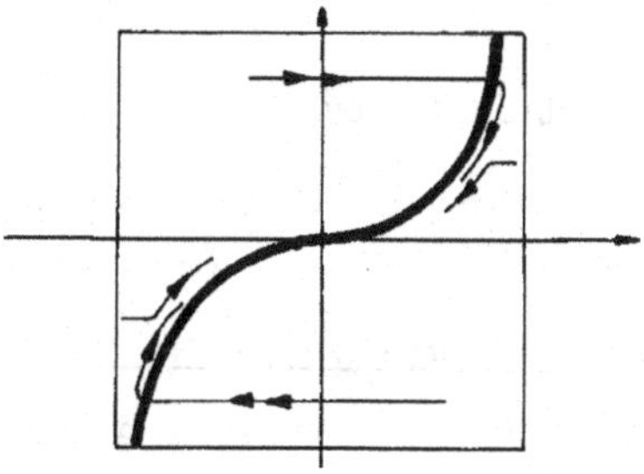

k > 0

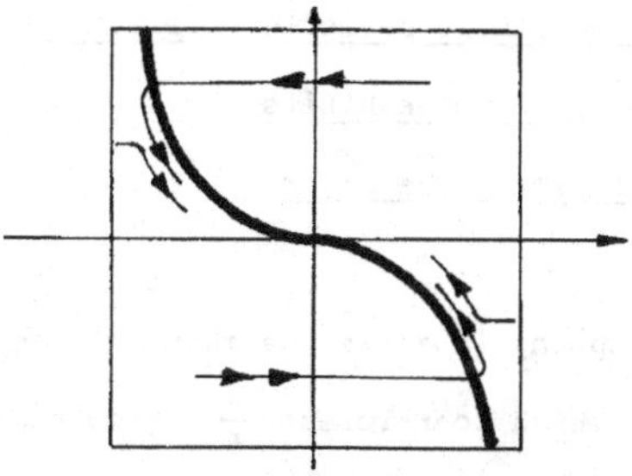

k< 0

(Note that the vector field is not pointing inward along the horizontal edges ; thus $K$ is not invariant under the flow.)

Thus every integral curve of III starting in the halo of $(0, 0)$ certainly remains in $K$ (of course, we know much more, but here we don't need it.

Now, consider a standard integral curve $\gamma$ of (II) and its transform $\Gamma$. Then $\Gamma(0)$ is in the halo of $0$, and $\Gamma(t) \in K$ for every $T>0$ ; thus for every $t>0$, $\gamma(t)$ is in the square of radius $\dfrac{1}{\varepsilon^n}$ and center $(0, 0)$ and so every standard integral curve of II is bounded (note that transfer yields also a standard bound for $\gamma$ ) and any one is bounded by transfer.

3) Exercises.

. Picture a positively invariant domain.

. Use the same trick to study boundedness of integral curves for non autonomous equations, for instance

$$x'' + f'(x)x' + g(t, x) = 0 .$$

. Consider the case where $f$ is not odd, that is for disymmetric viscosity.

. Try to find a classical proof following the same intuitive lines as the present one.

. What is the asymptotic behaviour as $t \to -\infty$ ?

4) Further applications can be found in [T] .

---

Lesson 7

RELAXATION OSCILLATIONS IN VAN DER POL'S EQUATION

Problem. Describe the asymptotic behaviour of integral curves in Van der Pol equation $\varepsilon x'' + (x^2 - 1)x' + x = 0$ as $\varepsilon$ tends to $0$ $(\varepsilon > 0)$. What is their relation with the solutions of the reduced equation $(x^2 - 1)x' + x = 0$ ?

THEOREM. For $\varepsilon$ little enough, there is an unique stable periodic oscillation which attracts quickly all other orbits. This oscillation exhibits two slow phases separated by two quick jumps only perceptible in a time of order $\varepsilon$ .

Comments. 0) Again we have as physical model a spring in a viscous thin cylinder, but force and viscosity are both reinforced by a high coefficient $\frac{1}{\varepsilon}$ . Also electrical systems (charge and discharge of a condenser) may lead to the present problem ; of course $x^2 - 1$ may be replaced by $F(x)$ without introducing much new difficulties in our discussion.

Notice that Van der Pol's equation may be considered as a qualitative model for heartbreaking, an important chapter of biology, indeed...

1) By Lienard's transformation, we get system

$$(1) \qquad \begin{cases} \varepsilon x' = y - \dfrac{x^3}{3} + x \\ y' = - x \end{cases}$$

and the cubic $\Gamma$ of equation $y = \frac{x^3}{3} - x$ plays a central part.
Let J. HAAG (Ann. Scient. E.N.S. 60 (1943), p. 35-111) tell us what is going on :
" Si $\varepsilon$ tend vers $0$, l'équation devient à la limite $(y - \frac{x^3}{3} + x)dy = 0$. La trajectoire limite est donc nécessairement constituée par des segments horizontaux et des arcs de $\Gamma$ ..."
This inference seemed quite an evidence in the first papers on the subject ; nowadays it has to be proved, and proofs in singular perturbations are often cumbersome : you have to approximate in successive steps with hard trouble whenever there are "corners" and corners are precisely what is important here !

2) With the help of lesson 5, the behaviour of the integral curves starting at a finite point $(x_0 , y_0) \neq (0 , 0)$ is easy to describe, because Van der Pol's equation is standard (in the general case of an $F(x)$, reduce the problem to standard $F$, by transfer).
Thus put $\varepsilon \sim 0$ and consider the corresponding slow-fast flow in the $(x , y)$ - plane. Using the symbols of lesson 5, we get the following description, which states precisely what J. HAAG had in mind.
Thus, after some finite time ( $\sim 0$ if $(x_0 , y_0)$ is not in the halo of $(0 , 0)$ ), the integral curve starting at $(x_0 , y_0)$ moves in the halo of the closed curve $a\,b\,c\,d$. The time spent along the horizontals $(d\,a)$ and $(b\,c)$ is infinitesimal, while along the arcs $(a\,b)$ and $(c\,d)$ of $\Gamma$, the time spent is infinitely close to $-\int_{x_a}^{x_b} \frac{dy}{x} \sim \int_1^2 \frac{x^2 - 1}{x}\, dx = \frac{3}{2} - \text{Log } 2$ and the total time along a "nearly oscillation" is $\sim 3 - 2 \log 2$.
Moreover, the quick "discharges" need a time which is less than any $\lambda\varepsilon$ with $\lambda$

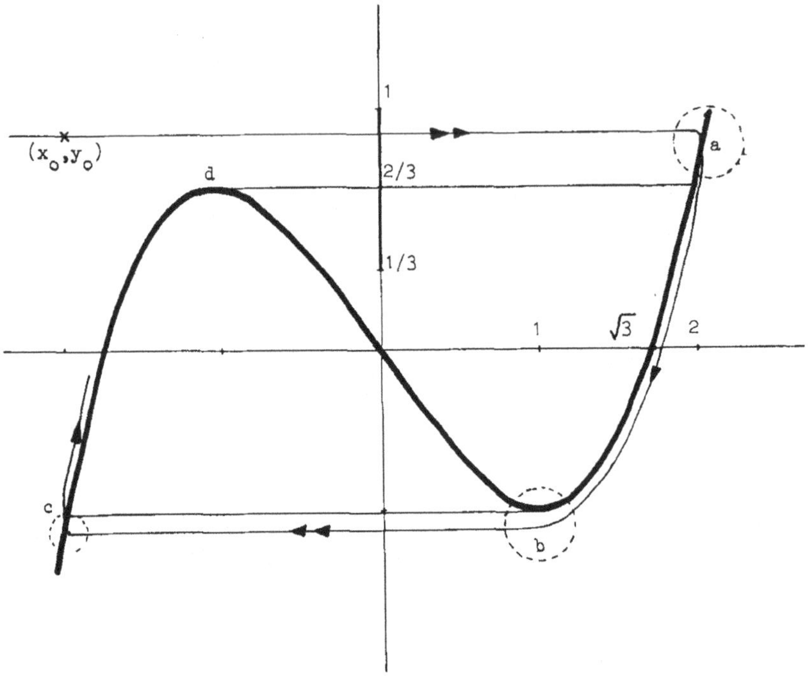


infinitely large. Hence the error in the time estimate above along an eventual closed orbit is of order $\varepsilon$ at most.

3) Consider the standard segment $[\frac{1}{3}, 1] = D$ on the y-axis. Every integral curve starting on $D$ turns once around the origin and meets again $D$ near $(0, \frac{2}{3})$. This yields a continuous mapping $\varphi : D \longrightarrow D$, which, following Brouwer, has a fixed point in the halo of $\frac{2}{3}$. Hence there exists a periodic oscillation, as expected above (alternative proof : use Poincaré-Bendixon). Moreover, between two eventual fixed points, $\varphi$ is monotoneous (no crossing of integral curves) and this implies that our oscillations are limit cycles. As the stability index $\int_0^T (1 - x^2)\, dt$ of any closed orbit $\gamma$ of period $T$ is $< 0$ (in the halo of $a\,b\,c\,d$, this is clear ; out of the finite plane also, for $1 - x^2$ would be $< 0$.), uniqueness and stability of the limit cycles follows, which proves the expected theorem, after some transfer.

4) A. TROESCH and E. URLACHER [T.U.] have studied the phenomenon in more details, using slow-fast flows which appear in successive changes of variable ; they look with these "microscopes" how the integral curves move along the arcs (a b) and (c d) . Refering the reader to the original papers for detailed proofs, we give only two examples :

- along (a b) and (c d) and outside of the halos of the end points, every integral curve has its derivatives at any standard order infinitely close to the derivatives of the cubic at the corresponding shadows.

This is a non-crackling phenomenon, which means that V.d.P. equation is well oiled. Of course, this is important for a heart breaking model !

- what is going on in the halo of b (or d ), as an integral curve moves away ?

Accurate changes of variables show that integral curves move between the two curves $\Gamma_1$ and $\Gamma_2$ of respective equations

$$y = \frac{x^3}{3} - x + \frac{\varepsilon x}{1-x^2} \quad \text{and} \quad y = \frac{x^3}{3} - x + \frac{\varepsilon x}{1-x^2}\left(\frac{1}{1-\frac{\varepsilon(1+x^2)}{(1-x^2)^3}}\right)$$

as long as $x(t) > 1$ .

A little computation gives a minimum $m_o = -\frac{2}{3} + k\varepsilon^{2/3}$ at $x_o = 1 + k'\varepsilon$ for $\Gamma_2$, with $k$ and $k'$ finite, not $\sim 0$ (precisely $k \simeq -0{,}0957$ and $k' \simeq 0{,}423$ ).

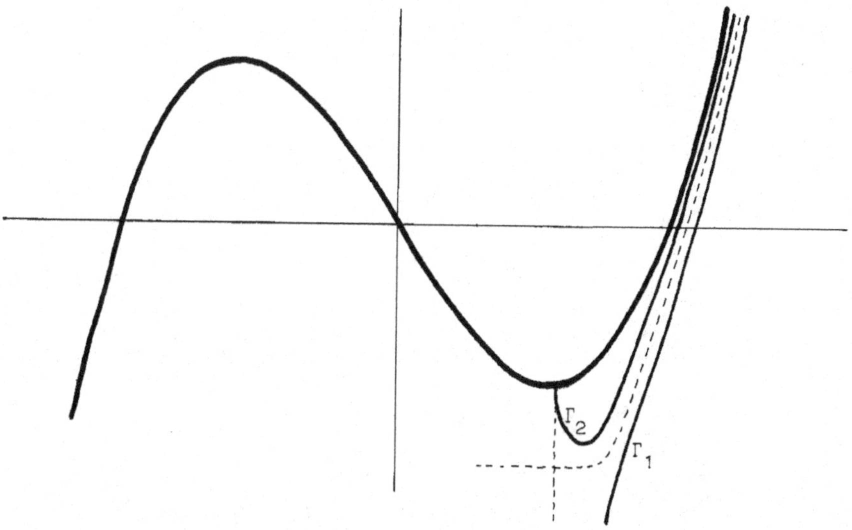

Thus the curve $\Gamma_2$ is a "spring-board" which gives to the moving point enough energy to get an infinitely large speed before learning the halo of b ; precisely under $(x_o , m_o)$ , the x-component of the speed is of order $\varepsilon^{-1/3}$ .

5) Exercises.

• Translate the results above in "$\varepsilon - \delta$"-words and imagine a classical proof.

• Study the more general equation $\varepsilon\ddot{x} + f(x)\dot{x} + x = 0$ along the same lines (see [U]).

• Study the slow-fast flow associated with such an equation in the usual phase-plane (it is a "vertical" flow, of course). Compare both pictures. Try other changes of variables, than Lienard's.

# Lesson 8

## CANARDS

Problem. How does the limit cycle disappear in equation

$$(E_a) \quad \begin{cases} \varepsilon \ddot{x} + (x^2 - 1)\dot{x} + x = a \\ \varepsilon > 0 \ , \quad \varepsilon \ \text{little enough} \end{cases}$$

when the parameter $a$ crosses the value $1$ $(a \geq 0)$ .

THEOREM. For $\varepsilon \sim 0$ and $a$ finite, the vector field $Z_a$ of components $\left(\frac{1}{\varepsilon}(y - \frac{x^3}{3} + x) \ , \ a - x\right)$ in the Lienard plane has a unique limit cycle for $a < 1$ and none for $a \geq 1$ . Whenever $a < 1$ , $a$ not $\sim 1$ , the cycle has the same shadow as for $(E_o)$ ; whenever $a \sim 1$ , the cycle shrinks down continuously to an infinitesimal cycle, through "canard-cycles".

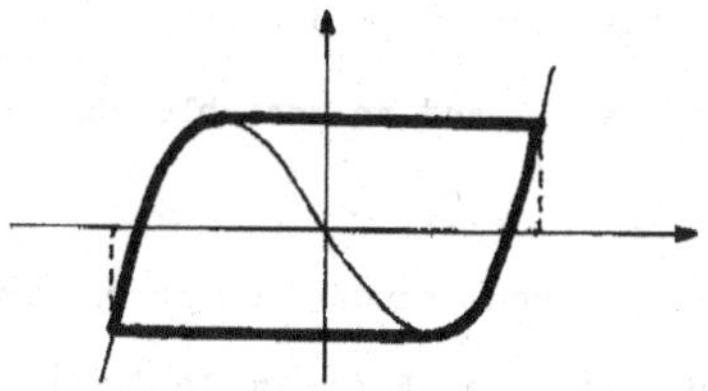

shadow of a large cycle

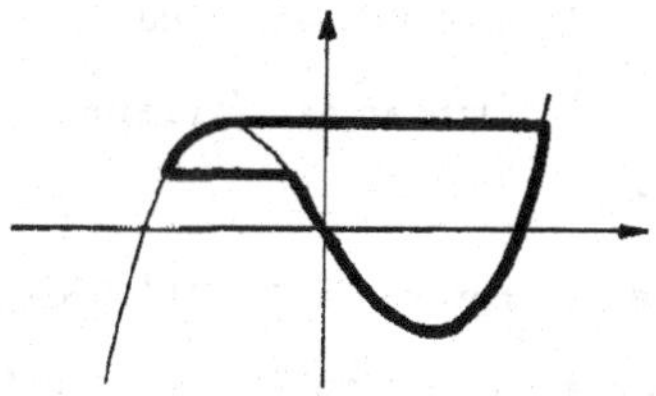

shadow of a canard ( with head )

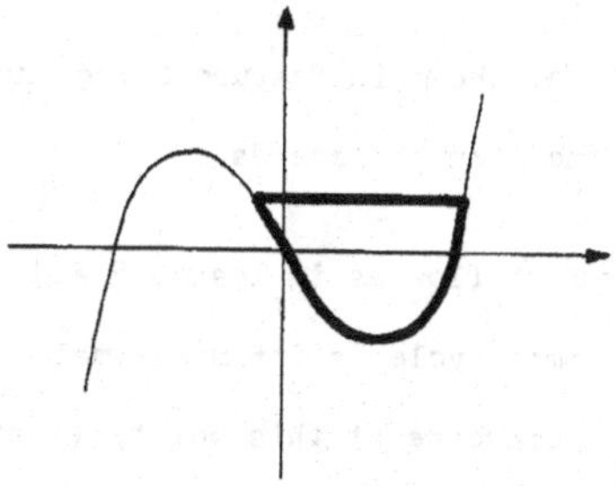

shadow of a canard ( without head )

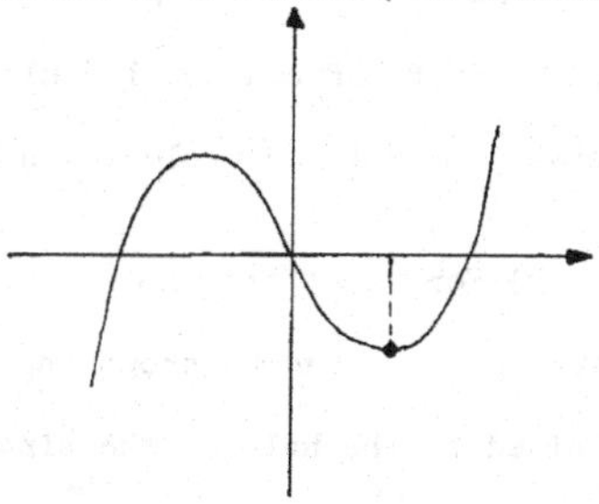

shadow of an infinitesimal cycle

Comments. 0) Uninformed people may be interested in "canards" (ducks in english) only from gastronomical point of view (usually "à l'orange"). Here we get only canards-on-the-paper whenever we observe the transition between large cycles as in Van der Pol's equation $E_o$ and the case without cycle $a>1$ ; yet the result is not unpleasing to the mathematical taste...

Indeed, in this two parameters singular perturbation problem, it is quite impossible to use the classical methods of asymptotic expansions, all the more that the phenomenon has first to be expected.

Moreover "discontinuous methods" (where $\varepsilon = 0^+$ ) like "constraint equations" forget the canard-cycles and only see a catastrophe for $a = 1$ (see F. TAKENS, Implicit differential equations : some open problems).

A close analysis of the problem using N.S.A. easily describes the actual transition. We only discuss the main points, refering to the original papers for details (see [B.C.D.D.]).

At this stage, it is no longer useful to translate the results in classical words, for the intuitive meaning is clear.

1) An elementary study of the field $Z_a$ at the singular point $(a, \frac{a^3}{3} - a)$ (linear part) yields a stable mode for $a>1$ and an unstable mode for $a<1$ .

This leads to foresee a Hopf bifurcation for $a = 1$ . The non-standard point of view gives here a quite complete description of the shape and size of the cycle during the quick transition.

In fact, the Hopf bifurcation is only the ultime stage, the main feature being the deformation from a large cycle to an infinitesimal one through canards.

2) For $a>1$ and not $\sim 1$, we get a slow-fast flow as in lesson 5 and the feature of the shadows shows that any eventual limit cycle is infinitesimal and contained in the halo of the singular point. A microscope at this point, (that is an homothety of infinitely large power) shows that such a cycle cannot occur (see fig. 1). Thus $Z_a$ has no limit cycle.

From the permanence principle, it follows that this is true until some $a \sim 1$ , $a>1$ .

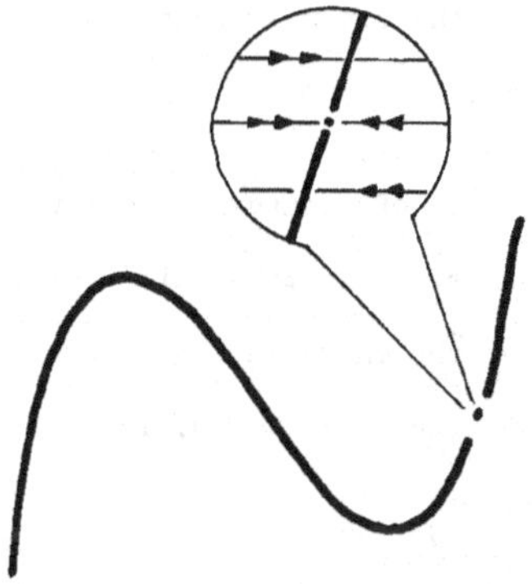

Fig.1 The singular point through the microscope :

$X=(x-a)/\delta$ , $U=(u-a^3/3+a)$

with $\delta\sim0$

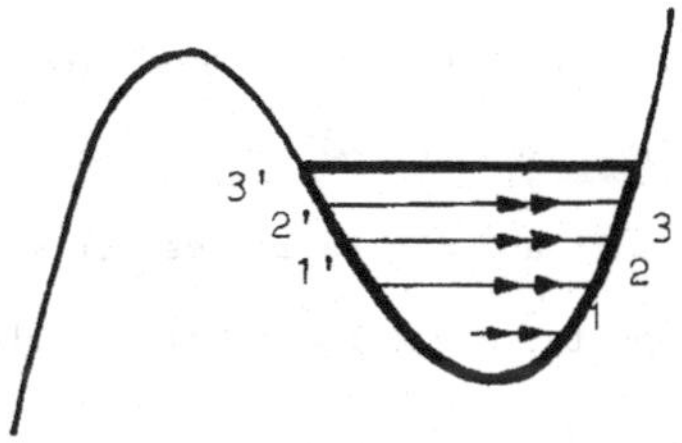


Fig.3

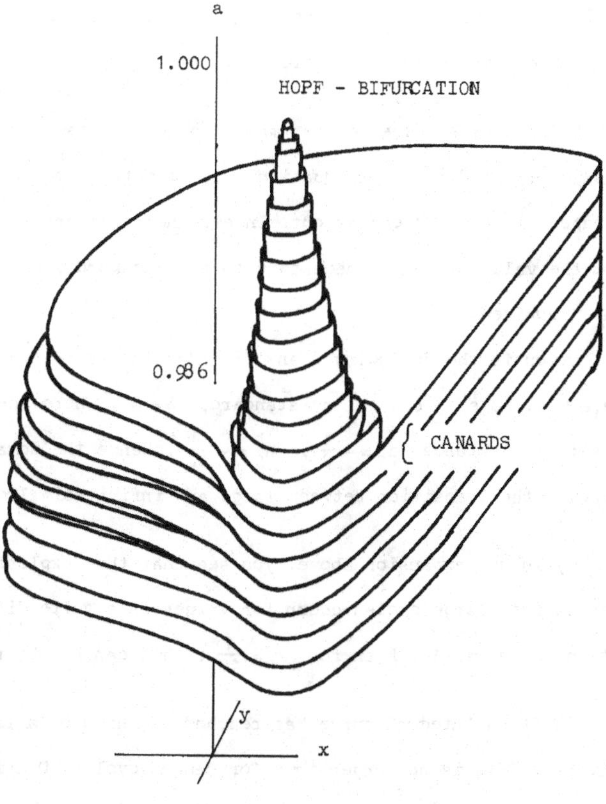


Fig.2

Now, it is easy to prove the asymptotic stability of the singular point for $a=1$. Hence even for $a \sim 1$, $a>1$ there is no limit cycle.

3) For $a<1$ and $a$ not $\sim 1$, the existence of a large cycle may be proved as in lesson 7 for equation $E_0$. By permanence, this cycle still remains (with the same shadow) until some $a \sim 1$. As before, the characteristic exponent $I$ of such a cycle satisfies $\varepsilon I \sim \int_0^T (x^2-1)dt < 0$ (and not $\sim 0$). Thus $I$ is $< 0$ and the cycle is asymptotically stable and unique.

4) The existence of canard cycles follows from (2) and (3), due to the continuity of the integral curves of $Z_a$ with respect to $a$. The important property of a canard is that it goes along a repulsing arc of the cubic, whereas large cycles behave as in Van der Pol's equation : only attractive arcs are used. In case the singular point is infinitely close to the lower vertice of the cubic, i.e. $a \sim 1$, it is possible for some values of $a$ that the cycle climbs along the foreside of the cubic near the vertice.

5) For which values of $a$ may we observe a canard ? If you fix a value of $\varepsilon$, say $\frac{1}{100}$ and put your equation in a computer, you have no chance to see a canard shaped limit cycle coming out. Indeed, nearly a precision up to the 60th digit for the value of $a$ is necessary to get a canard ; for $\varepsilon = \frac{1}{20}$, only 12 digits are required...

The explication is the following : any canard value is of type $a = 1 + a_1\varepsilon + \ldots + a_n\varepsilon^n + \delta\,\varepsilon^n$, $n$ standard, $\delta \sim 0$, where the $a_i$ are numerical constants ; for instance $a_1 = -\frac{1}{8}$, $a_2 = -\frac{3}{32}$. Hence the domain of all canards is very narrow : the transition between large and infinitesimal cycles is very quick.

6) In the expansion above, you see that the "explosion" of the infinitesimal cycles into large cycles occur for values of a quite different from the Hopf bifurcation value $a=1$. Even for $\varepsilon = \frac{1}{10}$, this can be observed (see fig. 2).

7) If an integral curve has reached the halo of a large cycle, it never will leave it. This is no longer true for canard cycles. On fig. 3, you see the

shadow of an integral curve tending to a canard. Whenever this shadow reaches the shadow of the limit cycle at $M$ , it leaves it at $M'$ . It is possible to prove that $M \longrightarrow M'$ is a functionnal relation. Precisely the $x$-components of $M$ and $M'$ are related by $P(x) = P(x')$ , where $P(u) = \dfrac{u^4}{4} + \dfrac{u^3}{3} - \dfrac{u^2}{2} - u$ .

### 8) Exercises.

- Observe the integral curves in the phase plane $(x , v = \overset{'}{x})$ , and in the stretched phase plane $(x , u = \varepsilon \overset{'}{x})$ . What is the shadow of the canards ?

- Prove that the period of a large cycle is (if $a$ not $\sim 1$) :
$$T \sim 3 + (a^2 - 1) \log \frac{4 - a^2}{1 - a^2} .$$

- Prove that equation $\varepsilon \overset{''}{x} + (x^2 - \mu) \overset{'}{x} + x = 0$ has a cycle for $\mu > 0$ and none for $\mu \leq 0$ . Prove that this is a simple Hopf bifurcation (without canards).

### 9) Remark. It seems that a lot of generic bifurcations occur with ca-
nards, that is curves which partly go along slow attractive and slow repulsive arcs (see the papers $[Df]$ , $[Dm]$).

---

Lesson 9

# GEODESICS ON FLATTENED SURFACES AND THE BILLIARD BALL PROBLEM

Problem. Let $S$ be a convex and compact surface in $\mathbb{R}^3$ and $S_\varepsilon$ its image under the mapping $(x, y, z) \longrightarrow (x, y, \varepsilon z)$. Describe the asymptotic behaviour of $S_\varepsilon$'s geodesics as $\varepsilon$ tends to $0$, $\varepsilon > 0$.

THEOREM. Consider $S_0$ as a billiard table in the $(x, y)$-plane. Then for any closed and indestructible orbit $\gamma$ of the billiard ball problem for $S_0$ and any neighbourhood $V$ of $\gamma$ in $\mathbb{R}^2$, there exists an $\varepsilon_0 > 0$ such that for every $\varepsilon < \varepsilon_0$, $S_\varepsilon$ has a closed geodesic whose projection is contained in $V$.

Comments. 0) "In days of yore the earth was nearly flat, a dreadful abyss won over poor sailors off shore, down in the hell, from waves to Devil's fire..."
So tragic was the situation of seamen, that scribes specialized in asymptotics hardly worked about the shortest way to explore the hidden side and come safely back into the harbours refuge. Fortunately, in 1927, G.D. BIRKHOFF had the good idea :
"If the surface is now flattened to the form of a plane convex curve, the billiard ball problem results. But in this problem the formal side, usually so formidable in dynamics, almost completely disappears, and only the interesting qualitative questions need to be considered".
In other words, closed geodesics of the flattened surface (for the metric induced by the Euclidean structure of $\mathbb{R}^3$) have something to do with closed orbits of a billiard ball running on a two sided billiard table along straight lines, reflecting on the boundary as in a mirror and changing symbolically the running side at each reflection (see fig. 1).
For this problem, there is a well-oiled topological technic, whereas geodesics of the actual smooth surface are quite impossible to describe.

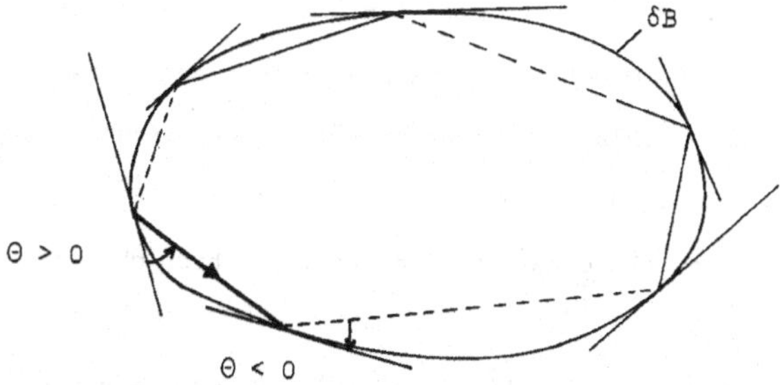


Fig.1

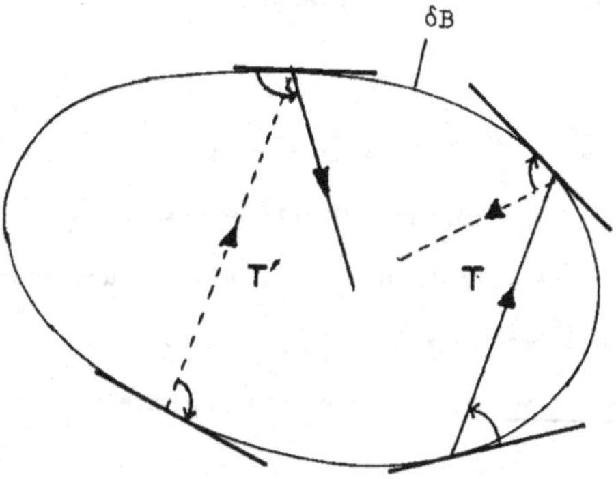


Fig.2

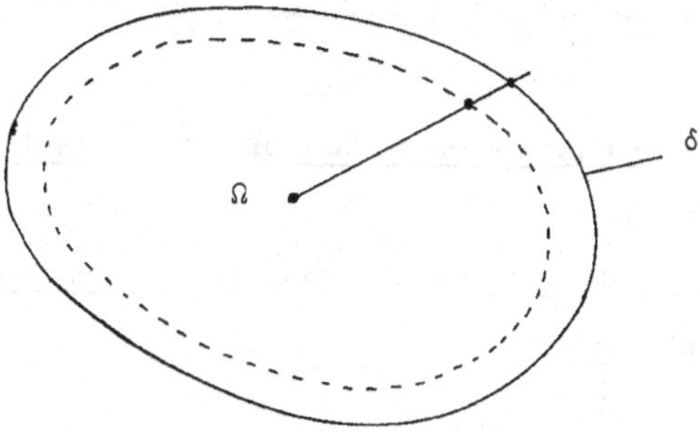


Fig.3

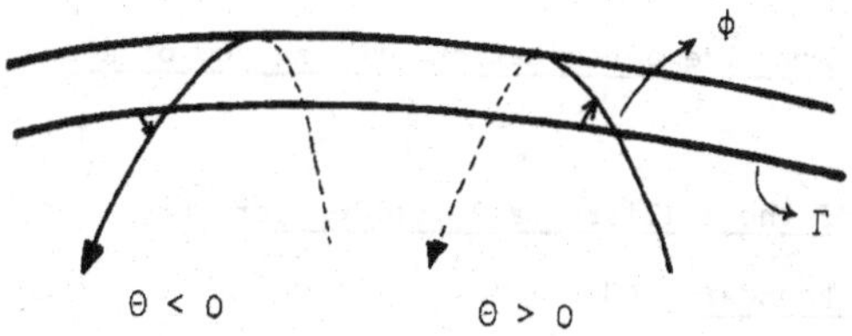


Fig. 4

However, Birkhoff didn't describe a precise relation between both problems. Of course, the question is no longer crucial, for meanwhile the earth became a sphere... Notice that this is a typical singular perturbation (or better singular deformation) problem : you have a (quite complicated) family of second order differential equations with a parameter $\varepsilon$ , i.e. the equation of geodesics, which changes in nature as $\varepsilon = 0$ .

A more general problem is, given a convex billiard table B , to describe the relation between geodesics of regular surfaces approaching the table and orbits of the billiard ball:

1) By transfer, we may assume B standard and describe the expected relation for any internal regular $C^2$ surface C whose shadow is B . Clearly the relation is weak for such any perturbation of B , and we must assume some restrictions on the quality of the approximation of B by C .

Precisely, call <u>a nice perturbation of</u> B any convex surface C in $\mathbb{R}^3$ with the following properties :

i) <u>The shadow of</u> C <u>is</u> B .

ii) <u>Outside of the halo of</u> $\partial B$ , <u>the tangent planes of</u> C <u>are nearly horizontal</u>.

iii) <u>Two infinitely near points in the halo of</u> $\partial B$ <u>may be joined by a geodesic with infinitesimal length</u>.

iv) <u>The curvature of</u> C <u>is "well-distributed"</u>, <u>i:e: its integral on any infinitesimal cell is infinitesimal</u>.

These conditions are of type $C^0$, $C^1$ and $C^2$ .

2) <u>Main Lemma</u>. <u>For any nice perturbation</u>, <u>the shadow of a geodesic arc of</u> C <u>with finite length is</u> :

- <u>either an orbit of the billiard ball problem for</u> B ,
- <u>or an arc of the boundary</u> $\partial B$ .

We give the essential steps of the proof ; each is an easy exercise.

a) Prove that outside of the halo of $\partial B$ , geodesics of C have infinitesimal curvature in $\mathbb{R}^3$.

b) Use the permanence principle to get a $\mu \sim 0$ such that outside of the $\mu$-neighbourhood of $\partial B$, geodesics of $C$ have infinitesimal curvature.

c) If $M \in \partial B$, prove that there exists an $\alpha \sim 0$ such that the homothetic image $\hat{C}$ of $C$ with center $M$ and power $\frac{1}{\alpha}$ has the following properties :

- it is nearly flat.
- its boundary is nearly straight.
- outside of the halo of $\partial B$, its geodesics have infinitesimal curvature.

d) If $P$ and $Q$ are two points on this new surface, prove that there exist a geodesic arc from $P$ to $Q$ whose shadow is an orbit of the "billiard with straight boundary" (construct a curve form $P$ to $Q$ whose shadow is such an orbit and prove that near it, some curve from $P$ to $Q$ has minimal length).

e) Use Gauss-Bonnet and assumption (iv) to prove that any geodesic arc of $\hat{C}$ is of the type above and deduce the lemma.

3) Let us recall Birkoff's procedure to get closed orbits in the billiard ball problem. First parametrize $\partial B$ by some diffeomorphism onto the circle $S^1$. Any reflection at some point of parameter $\varphi$ (mod $2\pi$) is described by the angle $\theta$ of the orbit with the tangent ; take $\theta \in ]0,\pi[$ if the ball leaves $\partial B$ on the upper side and $\theta \in ]-\pi,0[$ in the opposite case (see fig. 1).
Hence every reflection is paramatrized by a point of $S^1 \times ]0,\pi[$ or $S^1 \times ]-\pi,0[$, according to the order of the sides along the orbit.
Now we get an "upper next reflection mapping"

$$T : S^1 \times ]0,\pi[ \longrightarrow S^1 \times ]-\pi,0[$$

and a "lower next reflection mapping"

$$T' : S^1 \times ]-\pi,0[ \longrightarrow S^1 \times ]0,\pi[$$

(see fig. 2).
The closed orbits of the billiard ball are precisely the fixed points of some power $\varphi$ of $T' \circ T : S^1 \times ]0,\pi[ \longrightarrow S^1 \times ]0,\pi[$. If such a fixed point is non degenerate (i.e. the rank of $Id - \varphi^t$ is 2 at this point), then <u>any little perturbation of the mapping has a fixed point near the first one</u>. In particular, the fixed

point remains for any little change of $B$ . We call this an <u>indestructible closed orbit</u>.

4) Now consider a nice perturbation $C$ of a standard billiard table $B$ as in § 1. Let $\Omega$ be some standard inner point and $\partial\hat{B}$ the homothetic image of $\partial B$ with center $\Omega$ and power $k<1$, $k\sim 1$ (see fig. 3). We get a curve in the halo of $\partial\hat{B}$ , which is the projection of a curve $\Gamma$ on the "upper" part of $C$ . Clearly $\Gamma$ is in the halo of the curve on $C$ , which projects on $\partial B$ ; it is parametrized by construction.

To any geodesic $\lambda$ intersecting $\Gamma$ at a point of parameter $\varphi$ is associated the angle $\theta$ between $\Gamma$ and $\lambda$ with the accurate sign according to the side of incidence (see fig. 4). If $\theta$ is not $\sim 0$, the geodesic has as shadow an orbit of the billiard ball with the shadow of $(\varphi, \theta)$ as parameter (use the main lemma). Thus, as in the case of the billiard ball, we get for any $a>0$ , $a$ not $\sim 0$ , two "next intersection mappings"

$$\overline{T} : S^1 \times ]a, \pi - a[ \longrightarrow S^1 \times ]-\pi, 0[$$

and

$$\overline{T}' : S^1 \times ]-\pi + a, -a[ \longrightarrow S^1 \times ]0, \pi[ ;$$

a closed geodesic corresponds to a fixed point of some

$$(\overline{T}' \circ \overline{T})^n : S^1 \times ]a, \pi - a[ \longrightarrow S^1 \times ]0, \pi[ .$$

Due to the main lemma, for each $a>0$ , not $\sim 0$, and $n$ standard, $(T' \circ T)^n$ is infinitely close to $(\overline{T}' \circ \overline{T})^n$ on $S^1 \times ]a, \pi - a[$ . By the permanence principle, this property is true until some $\alpha_n \sim 0$ .

Now let $(\varphi_0, \theta_0)$ be some standard non degenerate fixed point of $(T' \circ T)^n$ ( $n$ standard). Then $\theta_0 \in ]\alpha_n, \pi - \alpha_n[$ ( convexity of $\partial B$ ) and $(\overline{T}' \circ \overline{T})^n$ has a fixed point $(\varphi_0', \theta_0') \sim (\varphi_0, \theta_0)$ . Thus <u>any indestructible closed orbit of the billiard ball is the shadow of some closed geodesic of</u> $C$ .

5) To get the expected theorem about $S_\varepsilon$ , first formulate it in its non-standard equivalent form (use easy transfers).

"<u>If</u> $S$ <u>is standard and</u> $\varepsilon\sim 0$ , <u>any indestructible</u>..."

Thus it only remains to prove that $S_\varepsilon$ is a nice perturbation of $B = S_0$ . We leave this to the reader : it is only classical differential geometry (use Gauss-Bonnet with geodesic triangles to prove (iv)).

For more details, and also a more general study of geodesics for nice approximations of cubes, polyedra, ..., see the original paper of J.L. CALLOT [C].

The reference for BIRKHOFF is "Dynamical systems, AMS coll. Publications, vol. 9, 1927".

# Lesson 10

## ASYMPTOTIC BEHAVIOUR IN BOUNDARY VALUE PROBLEMS WITH A SMALL PARAMETER

Problem. Describe the asymptotic behaviour of solutions in the two-point problems of type $P_\varepsilon$ $\begin{cases} f(\varepsilon, x, \dot{x}, \ddot{x}) = 0 \\ x(0) = A, \quad x(1) = B \text{ prescribed,} \\ \varepsilon > 0 \end{cases}$ as $\varepsilon$ tends to $0$.

Comments. 0) Here we enter in the fascinating world of "asymptotics", whose key words are jumps and layers.

Indeed, the main features which appear are quick jumps separated by slow relaxation motions, as $\varepsilon$ is little.

Sometimes jumps occur near the boundaries of the time interval, and classical technics are relatively efficient in dealing with such boundary layers. However in the last decade, people became more and more busy about problems with interior layers (also called "free" or "transition" or "shock" layers : the vocabulary in this "far-west" of applied mathematics is rather unformal !), that is quick jumps within the time interval, at places whose location is an a priori condition to use classical methods ; unfortunately this location is a hard problem and the only cases where some informations have been obtained are those which concern prescribed layer places (through accurate "stability conditions") or rather evident (although not easy to prove) locations.

If you throw an eye in the litterature, you get the feeling that "asymptotics" is really a big business, with a lot of computations, inequalities, a priori estimates, fine analysis, etc...

A whole book would be necessary to list all titles on this subject ; for a first contact, the reader will find in a paper by R.E. O'MALLEY "Topics in singular per-

turbations". Advances in Math. 2 (1968), pp. 365–470) a pleasant survey with a lot of references. Also see the six hundred references in the book of A.H. NAYFEH "Perturbation methods". Wiley-Interscience, New-York, 1973.
What makes people so heartful is the high importance of quick jumps for engineers, as well in hydro- aero- or magneto-dynamics as in oceanography, elasticity theory or predation modelising. Big effects for little causes – a typical character in singular perturbations, and a trouble in much technical systems !
Of course, A. Robinson had in the mind to use N.S.A. as a tool in layer problems, what is clear from his chapters on the boundary layer concepts in hydrodynamics and elasticity.

1) Although jumps and layers are geometric characters of asymptotic behaviours, classical papers never proceed with geometric arguments (see lesson 13 about an instructive exception). The classical trick is to compute a "formal solution" $\bar{x}_\varepsilon$ with expansions technics and then to prove its closeness to an actual solution $x_\varepsilon$ as $\varepsilon \to 0$, usually up to some order of $\varepsilon$. Traditions help to choose the accurate type of the terms in $\bar{x}$ (but in case of unknown free layers, traditions fail...), and then you prove on $\bar{x}_\varepsilon$ the expected behaviour of $x_\varepsilon$.
The main inconvenient of this procedure is that you must expect a priori some behaviour, and then try to get your formal solution $\bar{x}_\varepsilon$ of the wanted type. Of course, the closeness with an $x_\varepsilon$ is by no means trivial and sometimes there is no $x_\varepsilon$ close to $\bar{x}_\varepsilon$ (see lesson 13).

2) Now let us discuss the formulation of asymptotic behaviours. The main concept is boundary (or interior) layer character of thickeness order $\varepsilon^\nu$, which concerns a family $x_\varepsilon$ of solutions of $(P_\varepsilon)$.
A general definition of this concept is immaterial if you only get your behaviours via formal approximations : if you have in $\bar{x}_\varepsilon$ some term $\varphi(\frac{t-t_0}{\varepsilon^\nu})$ where $\varphi(u) \longrightarrow 0$ as $|u| \longrightarrow \infty$, you claim that the layer is of thickeness order $\varepsilon^\nu$.
However general definitions of typical behaviours should be available to unify somewhat the singular perturbation theory. Unfortunately, the classical language

is not suitable to formulate asymptotic behaviours easily.
Indeed, engineers ask "what is going on with the solutions of $P_\varepsilon$ as $\varepsilon$ is little" ; you try to answer in terms of limits : "as $\varepsilon$ tends to $0$ , the solutions tend to...", but in much cases, there is no uniform limit on the whole time interval. The problem is precisely to describe non uniformities... In any case you give your answer in terms of a family $x_\varepsilon(t)$ ; this is only valuable if there is some uniqueness argument or if several families of solutions can be clearly isolated from each other (notice that a general concept of "solution" as the set of all solutions for each $\varepsilon$ , as we had for polynomials in lesson 1 is meaningless for applied mathematics). However we should also describe very bad behaviours, like oscillating jumps with little period or unbounded solutions, where asymptotic developpements are lost...
Thus general definitions should avoid families depending on $\varepsilon$ . We want " $\varepsilon$-little enough" properties for <u>all</u> solutions and this is cumbersome.

3) With a non-standard mind, say within I.S.T., we may formulate the problem as follows.
First, transfer the question, that is assume all its constants (here $f$ , $A$ and $B$ ) to be standard. This is not a restriction, because any information which is true for all standard constants is true in the general case, due to the transfer principle.
Now, we know that all "$\varepsilon$-little enough" statements on $(P_\varepsilon)$ may be formulated as equivalent "for every $\varepsilon \sim 0$ " properties. The last are suggested by intuitive descriptions, and are quite natural to observe.
Thus we fix $\varepsilon \sim 0$ and look at all solutions of $P_\varepsilon$ ; any information which is true for all $\varepsilon \sim 0$ is an answer for our engineer (he has only to replace $\sim 0$ by "little").
For instance, if we have a standard function $\bar{x}(\varepsilon , t)$ and prove that for every $\varepsilon \sim 0$ , there is a solution $x(t)$ such that $\frac{x - \bar{x}(\varepsilon , t)}{\varepsilon^\nu}$ is finite, we claim that for $\varepsilon$ standard and little enough, there is a solution $x(t)$ such that $\left|\frac{x(t) - \bar{x}(\varepsilon , t)}{\varepsilon^\nu}\right| < M$ , where $M$ is any standard given number ; this transfers in-

to the classical answer. But to formulate this, we don't need N.S.A. ; we need it if we want to avoid to compute a priori an $\bar{x}$ ...

Consider a non standard function $x(t)$ on $[0,1]$ and an infinitesimal $\eta > 0$. Assume that $x(t)$ takes only finite values and call $w(t)$ its shadow. Thus, for every standard $t$, we have $x(t) \sim w(t)$. Nevertheless, $w(t)$ may be quite horrible and we have to describe different types of non-continuous shadows.

i) $w(t)$ is continuous. This is the case in regular perturbation theory. For instance

$$\begin{cases} \ddot{x} = \varepsilon \\ x(0) = A \\ x(1) = B \end{cases}$$

has an unique solution $x_\varepsilon(t) = A + (B - A - \frac{\varepsilon}{2})t + \varepsilon \frac{t^2}{2}$ and for $\varepsilon \sim 0$, $x_\varepsilon(t) \sim w(t) = A + (B-A)t$, which is the solution of the reduced problem $(\varepsilon = 0)$.

In classical words, $x_\varepsilon$ has $w$ as uniform limit on $[0,1]$.

Now a simple non-uniformity :

ii) $w(t)$ is continuous on $]0,1]$ and has a continuous extension $\bar{w}$ to $[0,1]$ such that $x(0) - \bar{w}(0)$ is not $\sim 0$.

Hence for every standard $t$, $t \neq 0$, we have $x(t) \sim \bar{w}(t)$. We call this situation boundary layer character at $0$.

Suppose that, as long as $\frac{t}{\eta}$ is infinitely large, we have $x(t) \sim \bar{w}(t)$ : Then we say that the layer is of thickeness order at most $\eta$ (it may be less than $\eta$).

Furthermore, if for every finite $\frac{t}{\eta}$, $x(t) - \bar{w}(t)$ is not $\sim 0$, we say that the thickeness order is $\eta$.

The prototype of this behaviour is the exponential decay $x(t) = \exp -\frac{t}{\eta}$, $\bar{w}(t) = 0$.

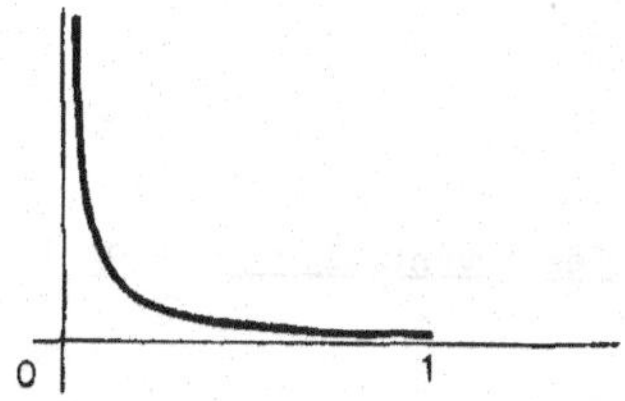


Clearly, we have a symmetric case with layer at time 1. The intuitive meaning of thickeness is that the jump near 0 is not perceptible for $\frac{t}{\eta}$ large, but is soon not neglectible for $\frac{t}{\eta}$ "not large".

Within N.S.A. this has a precise meaning ;

notice that between "large" and "not large" the distinction is not precise, whereas in N.S.A. it is precise, but shares with the intuitive concept the "softness" of the transition.

An important feature is that, due to the permanence principle, there exists a $t_o \sim 0$ such that $x(t) \sim \bar{w}(t)$ on $[t_o, 1]$, the jump occuring on $[0, t_o]$. This allows to study the behaviour of $x(t)$ in two steps : first on $[t_o, 1]$, and then on $[0, t_o]$, with different arguments, but without any "matching" problem.

iii) $w(t)$ is continuous on $[0, \ell[ \cup ]\ell, 1]$ with limits $w(\ell - 0)$ and $w(\ell + 0)$ such that at least two of the numbers $x(\ell)$, $w(\ell - 0)$, $w(\ell + 0)$ are not infinitely near, for some standard $\ell \in ]0, 1[$.

Thus, for every standard $t \neq \ell$, we have $x(t) \sim w(t)$. We call this transition (or inner, or free, ...) layer character at $\ell$. The word "free" recalls that the point $\ell$ is not given in problem $(P_\varepsilon)$ ; just here is the trouble, indeed !

Again define thickeness order $\eta$ using $\left|\frac{\ell - t}{\eta}\right|$ instead $\frac{t}{\eta}$. We have two prototypes (up to symmetries), according to the relative positions of $x(\ell)$, $w(\ell - 0)$, $w(\ell + 0)$. We only picture them, without writing functions.

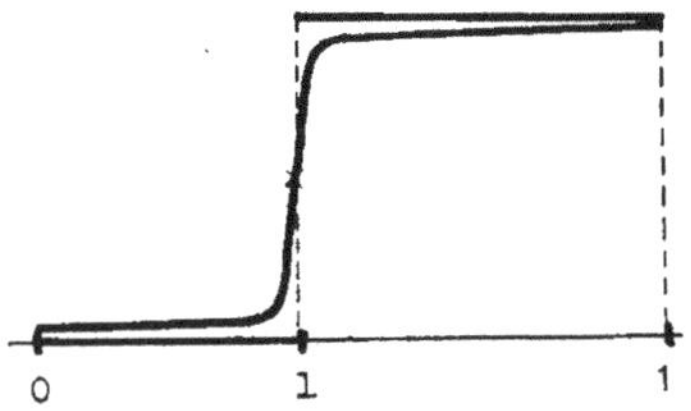


$w(\ell-0) < x(\ell) < w(\ell+0)$

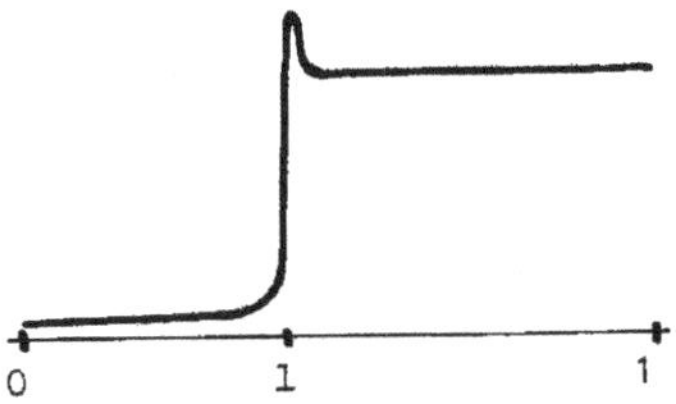


$w(\ell-0) < w(\ell+0) < x(\ell)$

The next case is a combination of (ii) and (iii).

iv) $w(t)$ has a finite number of discontinuities (eventually at 0 or 1), with limits (on both sides for the interior ones).

Here we have all kinds of layers and the thickeness orders may be different, although this is usually not the case in solutions of differential equations. Other behaviours may occur, but we have no name for them..

For instance, limits may fail at an isolated discontinuity, a rather complicated case. In lesson 13, we shall get the case of an infinite set of discontinuities, like $x(t) = \sin \frac{t}{\eta}$ ; there are oscillating jumps with infinitesimal period ! Very dangerous in a technical system, of course...

We also may forget the finiteness condition for $x(t)$ ; there is no shadow on $[0, 1]$ . A prototype is $x(t) = \frac{1}{\eta} \sin \frac{t}{\eta}$ . There is no vocabulary, for engineers don't expect unbounded solutions and mathematicians don't like to consider them...

As for classical equivalents of layer characters, we give only one example to make the comparison possible :

Consider a standard family $x(\varepsilon, t)$ of functions on $[0, 1]$ . The statement "for every $\varepsilon \sim 0$ , $x(\varepsilon, t)$ has boundary layer character at $0$ of thickeness order at most $\varepsilon^{\nu}$ and extended shadow $\overline{w}(t)$ " is equivalent (use successive transfers) to

" $\forall \lambda > 0, \ \exists R > 0, \ \exists \alpha > 0, \ \forall \varepsilon > 0, \ \forall t \in [0, 1]$ ,

$$(|\varepsilon| < \alpha \text{ and } |\frac{t}{\varepsilon^{\nu}}| > R) \Longrightarrow |x_{\varepsilon}(t) - \overline{w}(t)| < \lambda \text{ ".}$$

For thickeness precisely of order $\varepsilon^{\nu}$ , you have to complete the statement.

4) So long about formulations, but what about proofs ?

Clearly, our main tool is phase space observation in different stretched spaces, as we had for Van der Pol's equation, using the permanence principle to join slow motions and quick jumps in the study of the whole portrait of the integral curves. What is new here is the boundary value problem ; we have to predict the behaviours of these integral curves, whose projection on the x-axis goes from A to B between times 0 and 1 . In non pathological cases, the best of the time is spent along some solution of the reduced equation. Thus we try to detect the asymptotic behaviours of the actual solutions. As for asymptotic solutions, that is computable uniform approximations of actual solutions, we may very easily get in some cases as a by product an $\overline{x}$ with $x(t) - \overline{x}(t) \sim 0$ on $[0, 1]$ , but more precision on the approximation, like some order of $\varepsilon$ , needs more work. So, in the question of behaviours, we go further and easier than the classical technics, whereas about asymptotic developpements, we only avoid some computations, replacing inequalities by infinitesimal calculus. Indeed, we have the naive (and certainly dull) feeling that in asymptotics, the technic of developpements (which is nice but tiring) surreptitiously changed the problem : to get a developpement with high precision seems to be the fundamental question (but goes only in simple cases...). In other parts of mathematics, observations of the same kind may be done : the original problems often are reformulated in terms of a machinery which secretes its own exigences about the problem. The danger is that new and simpler approaches may be rejected, for they solve the old problem with a new mind and not the new problem with an old mind...

5) Let us end this discussion with a rather general information about non autonomous equations. Up to now, we had only to do with slow motions in phase planes along curves, for our equations were not time-depending. What about slow motions in a phase space $(t, x, \dot{x})$ ?

Consider equation $\varepsilon \ddot{x} = f(t, x, \ddot{x}, \varepsilon)$ with $f$ standard and $\varepsilon > 0$, infinitesimal. In the usual phase space we have system

$$(F) \qquad \begin{cases} \dot{x} = u \\ \dot{u} = \frac{1}{\varepsilon} f(t, x, u, \varepsilon) \\ \dot{t} = 1 \ . \end{cases}$$

Assume that $f$ is continuous.

Then we have the following

SLOW MOTION LEMMA. <u>Let</u> $\gamma(t) = (t, x(t), u(t))$ <u>be an integral curve of</u> (F) <u>on some standard time interval</u> $]\alpha, \beta[$ ; <u>assume that</u> $\gamma(t)$ <u>and</u> $\dot{\gamma}(t)$ <u>take finite values on</u> $]\alpha, \beta[$ . <u>Then the shadow</u> $w(t)$ <u>of</u> $x(t)$ <u>is a solution of the reduced equation</u> $f(t, w(t), \dot{w}(t), 0) = 0$ <u>on</u> $]\alpha, \beta[$ .

<u>Proof</u>. Let $t$ be standard, $t \in ]\alpha, \beta[$ . As $f$ is continuous, $\gamma$ and $\dot{\gamma}$ are continuous ; hence $\|\gamma\|$ and $\|\dot{\gamma}\|$ are bounded on any closed standard interval around $t$ by some standard constant $K$ (use their finiteness). Now we have, for

any standard h (little enough)

$$w(t_o+h)-w(t_o) \sim x(t_o+h) - x(t_o) = hu(t_o+\theta h) = (u(t_o)+\theta h\dot{u}(t_o+\overline{\theta}h)) ,$$

with $\theta, \overline{\theta}$ in $]0,1[$ . Hence $\left|\frac{x(t_o+h)-x(t_o)}{h} - u(t_o)\right| < hK$ and also $\left|\frac{w(t_o+h)-w(t_o)}{h} - ({}^\circ u)(t_o)\right| < hK$ , which proves that $w$ has ${}^\circ u$ as derivative at any standard point, hence by transfer at any point in $]\alpha, \beta[$ .

To end the proof, write $f(t, w(t), \dot{w}(t), 0) \sim f(t, x(t), u(t), \varepsilon) = \varepsilon\dot{u}(t)$ for any standard $t$ ; as $\dot{u}(t)$ is finite, we get $f(t, w(t), \dot{w}(t), 0) \sim 0$ , hence $f(t, w(t), \dot{w}(t), 0) = 0$ by standardness.

Notice that this lemma allows us to generalize somewhat the results of Lesson 5: the surface of equation $f(t, x, u, 0) = 0$ will play an essential part in the description of the integral curves starting at a finite point. In the next lessons, we study some typical problems with $f$ mainly of type $a(t, x)u + b(t, x)$ , the crucial point being the sign of $a(t, x)$ . The last lesson opens the door on the case of partial differential equations, where everything is really hard to work out !

# Lesson 11

## A SEMI-LINEAR PROBLEM WITH BOUNDARY LAYER

Problem. Describe the asymptotic behaviour of the solutions, as $\varepsilon$ tends to $0$, of the semi-linear two-point boundary value problem

$$(P_\varepsilon)\quad \begin{cases} \varepsilon \ddot{x} = a(t\,,x)(\dot{x} - b(t\,,x))\,, & 0<t<1 \\ x(0)=A\,,\quad x(1)=B & \text{prescribed} \\ a(t\,,x) \geq k > 0\,, & \varepsilon > 0\,. \end{cases}$$

THEOREM. Assume $a$ and $b$ Lipschitzian, bounded by a constant $M$ on $[0\,,1]\times\mathbb{R}$. Then for $\varepsilon$ little enough, $(P_\varepsilon)$ has a solution ; any family $x_\varepsilon(t)$ of solutions is bounded and has boundary layer character at time 1 of thickeness order $\varepsilon$. More precisely, if $w(t)$ is the solution of the reduced problem $\begin{cases}\dot{w}=b(t\,,w)\\ w(0)=A\end{cases}$, then on any closed interval $[0\,,s]\subset[0\,,1[$, one has the uniform approximates $x_\varepsilon(t)-w(t) = O(\varepsilon)$ and $\dot{x}_\varepsilon(t)-\dot{w}(t)=O(\varepsilon)$.

Comments. 0) In this problem, condition $a\geq k>0$ is an essential bolt to avoid free layers (the case $a\leq k<0$ has the same behaviour, with layer at $0$ instead $1$ ). Classical technics about it are well known (e.g. see A.B. VASIL'EVA : "Asymptotic behaviour of solutions to certain problems involving non linear differential equations containing a small parameter multiplying the highest derivatives". Russian Math. Surveys 18 (1963), pp. 13-84.), although not so easy to work out, especially if $a$ and $b$ are not smooth enough, since one needs at least second order terms in a formal solution to prove that the first order development is an approximate of an actual solution (and then deduce the layer character, observing the formal solution). Maliciously we ask about the behaviour whenever $a$ and $b$ are only Lipschitzian, just to insure uniqueness and continuity of the flow (see lesson III.7). Of course, we don't get an uniform approximate up to order $\varepsilon$, as wanted by the traditions, however, we get the essential geometric characters, including an uniform approximate on $[0\,,1]$. The boundedness of $b$ avoids unbounded families of solutions, and together with the boundedness of $a$, insures the existence of the flow until time $1$.

1) Following lesson 10, first transfer the statement, i.e. assume $a, b, A, B, k, M$ standard and fix $\varepsilon \sim 0$. In the usual phase space we get system

$$(I) \qquad \begin{cases} \dot{x} = u \\ \dot{u} = \frac{a}{\varepsilon}(u - b) \\ \dot{t} = 1 \end{cases}$$

and look for integral curves starting on the vertical $V_A$ of $(0, A)$, reaching the vertical $V_B$ of $(1, B)$ (see fig. 1).

Call $S$ the standard surface of equation $u = b(t, x)$ and $w$ the standard solution of the reduced problem $\begin{cases} \dot{w} = b(t, w) \\ w(0) = A \end{cases}$, where the boundary condition at time $1$ is lost. Put $w(1) = \overline{B}$; in general, we have $\overline{B} \neq B$ and $w$ cannot be an approximating solution of $(P_\varepsilon)$ on the whole time interval. We assume $\overline{B} < B$ (hence $\overline{B}$ not $\sim B$, since both are standard); the case $\overline{B} > B$ is similar.

Now observe an integral curve $\gamma(t) = (t, x(t), u(t))$ on $[0, 1]$, starting at $(t_o, x_o, u_o)$.

LEMMA 1. $u_o \geq 2M$ __implies__ $u(t) \geq 2M$ __and__ $\dot{u} > 0$ __on__ $[t_o, 1]$ __and__ $u_o \leq -2M$ __implies__ $u(t) \leq -2M$ __and__ $\dot{u} < 0$.

__Proof__. If $u_o > M$ and $u(t) \leq M$ for some $t > t_o$, $\dot{u}$ must vanish somewhere on $]t_o, t]$, since $\dot{u}(t_o) > 0$. Call $s$ the first time such that $\dot{u}(s) = 0$. Then $u(s) = b(s, x(s)) \leq M$ and for some $\xi \in ]t_o, s[$, $u(s) = u_o + (s - t_o)\dot{u}(\xi) > u_o \geq 2M$, which is a contradiction.

2) __Existence and finiteness of solutions__. Start on $V_A$, with $u_o \geq 2M$; by lemma 1, $\dot{u}(t) > \frac{k}{\varepsilon}(u(t) - b(t, x(t))) \geq \frac{k}{\varepsilon} M$ is infinitely large on $[0, 1]$. Hence $u(t) = u_o + t\dot{u}(\xi)$ is infinitely large for any $t$ not $\sim 0$. It follows that $x(1)$ is infinitely large. Starting with $u_o < -2M$, we get similarly $x(1)$ infinitely large $< 0$.

By continuity, there is an $u_o$ such that $x(1) = B$, since $B$ is standard. __Thus problem__ $(P_\varepsilon)$ __has a solution__.

Consider some solution $\gamma(t)=(t, x(t), w(t))$ of $P_\varepsilon$. If $x(t)$ is infinitely large, there is some $s\in]0,t[$ with $u(s)$ infinitely large. Hence by Lemma 1, $u$ is positive on $[s,1]$, which implies $x(1)>x(t)$ ; this is not possible, since $x(1)=B$ is standard. Similar argument for $x(t)$ infitely large negative. Thus <u>Any solution of</u> $(P_\varepsilon)$ <u>is finitely valued.</u>

From now on we observe a solution $\gamma(t)$ of $P_\varepsilon$.

LEMMA 2. <u>On any interval</u> $[0,\delta]\subset[0,1]$, $\delta$ <u>not</u> $\sim 1$, <u>one has</u> $u(t)\sim b(t, x(t))$.

<u>Proof</u>. Suppose that for some $t_o\in[0,\delta]$, $u(t_o)>b(t_o, x(t_o))$, not $\sim$. If $u(t_o)<2M$, the curve $\gamma$ is nearly vertical with high speed ; hence $u(t)$ reaches $2M$ after an infinitesimal time and as $x(t_o)$ is finite, we conclude as in §(2) that $x(1)$ is infinitely large, which is a contradiction. Similar proof for the opposite inequality.

Now consider $z(t)=x(t)-w(t)$ on $[0,\delta]$. It is the unique solution of

$$\begin{cases} \dot{z}(t) = u(t) - b(t, x(t)) + b(t, w(t)+z(t)) - b(t, w(t)) \\ z(0)=0 \end{cases}$$

by lemma 2, $z(t)$ is infinitely close to the unique solution of

$$\begin{cases} \dot{p}(t) = b(t, w(t)+p(t)) - b(t, w(t)) \\ p(0)=0 \end{cases}$$

This solution is clearly $p(t)=0$. Thus

<u>As long as</u> $t$ <u>is not</u> $\sim 1$, <u>we have</u> $x(t)\sim w(t)$ <u>and hence</u> $u(t)\sim b(t, w(t))$.

In other words, <u>both</u> $x(t)$ <u>and</u> $u(t)$ <u>have boundary layer character at time</u> 1.

Moreover, the approximations are permanent until some $t_o\sim 1$.

From continuity and standardness of $w$, we infer that $w(t_o)\sim w(1)=\bar{B}$ and $u(t_o)\sim b(1,\bar{B})$.

Now, as long as $u(t)$ is finite, $x(t)$ remains $\sim\bar{B}$ ; hence, as $B>\bar{B}$, not $\sim$, infinitely large values of $u$ occur on $[t_o,1]$ and by permanence, <u>there is some</u> $t_1\in]t_o,1[$ <u>such that</u> $u(t_1)$ <u>is infinitely large and</u> $x(t)\sim\bar{B}$ <u>on</u> $[t_o,t_1]$.

Taking eventually an intermediate value, we may assume $\varepsilon u(t_1)\sim 0$. From lemma 1, we know that on $[t_1,1]$, <u>both</u> $u(t)$ <u>and</u> $x(t)$ <u>are growing</u>. <u>Hence</u> $\dot{x}(t)=u(t)$ <u>is infinitely large on this interval.</u>

4) To see what is going on after time $t_1$, we use the "telescope" $y = \varepsilon u$ and observe the integral curves in the stretched phase space. We get system

$$(II)\quad \begin{cases} \dot{x} = \dfrac{y}{\varepsilon} \\ \dot{y} = a(\dfrac{y}{\varepsilon} - b) \\ \dot{t} = 1\ , \quad t \geq t_1 \end{cases}$$

Hence $\dfrac{\dot{y}(t)}{\dot{x}(t)} = a(t, x(t)) - \dfrac{a(t, x(t))\, b(t, x(t))}{\dot{x}(t)} \sim a(t, x(t)) \sim a(1, x(t))$,

since $|ab| \leq M^2$ (with $M$ standard).

Thus $\frac{k}{2}\dot{x}(t) < \dot{y}(t) < 2M\dot{x}(t)$ and $\frac{k}{2}(x(t) - x(t_1)) < y(t) - y(t_1) < 2M(x(t) - x(t_1))$.

As $x(t_1) \sim \overline{B}$ and $y(t_1) \sim 0$, we infer that $y(t) \sim 0$ <u>is equivalent to</u> $x(t) \sim \overline{B}$ and also that $y(t)$ <u>is finite</u> on $[t_1, 1]$ (recall that $x(t)$ is finite).

Now we have $x(1) - x(t) = \dfrac{1-t}{\varepsilon} y(\xi) > \dfrac{1-t}{\varepsilon} y(t)$ ; hence <u>if</u> $\dfrac{1-t}{\varepsilon}$ <u>is infinitely large</u>, $y(t)$ <u>is</u> $\sim 0$ <u>and</u> $x(t) \sim \overline{B}$, <u>i.e. the boundary layers are of thickeness order at most</u> $\varepsilon$ :

5) Put $T = \dfrac{1-t}{\varepsilon}$ and $X(T) = x(1 - \varepsilon T)$, $Y(T) = y(1 - \varepsilon T)$. From (II) we get

$$(III)\quad \begin{cases} \dot{X} = -Y \\ \dot{Y} = -a(Y - \varepsilon b) \\ \dot{T} = 1 \end{cases}$$

and our solution of $(P_\varepsilon)$ satisfies $X(0) = B$, $Y(0) = y(1) \sim {}^\circ(y(1))$. Hence, for any finite $T$, we have $X(T) \sim \overline{X}(T)$ and $Y(T) \sim \overline{Y}(T)$, where $(\overline{X}, \overline{Y})$ is the solution of

$$(IV)\quad \begin{cases} \dot{\overline{X}} = -\overline{Y} \\ \dot{\overline{Y}} = -a(1, \overline{X})\overline{Y} \\ \dot{T} = 1 \end{cases} \quad \text{with initial conditions} \quad \begin{cases} \overline{X}(0) = B \\ \overline{Y}(0) = {}^\circ(y(1)) \end{cases}.$$

This standard curve in the $(X, Y)$-plane tends to a standard singular point $(B', 0)$ and for any finite $T$, $(\overline{X}(T), \overline{Y}(T))$ is not infinitely close to this point (it is close to $(\overline{X}({}^\circ T), \overline{Y}({}^\circ T))$, which is standard, hence not close to the singular point).

From this, we infer that the corresponding curve $(t, \overline{x}(t), \overline{y}(t))$ is infinitely

close to $(t, x(t), y(t))$ for $\frac{1-t}{\varepsilon}$ finite. This property is permanent until some $t_2 \sim 1$ with $\frac{1-t_2}{\varepsilon}$ infinitely large. But then, we know that $x(t_2) \sim \bar{B}$ and $y(t_2) \sim 0$ ; hence $B' = \bar{B}$ , which determines $\bar{Y}(0)$ as the unique standard point such that the integral curve of IV starting there has $(\bar{B}, 0)$ as limit (see fig. 2).
Moreover, for $\frac{1-t}{\varepsilon}$ finite, $x(t)$ is not $\sim \bar{B}$ and $y(t)$ not $\sim 0$ .
Thus the boundary layers are of thickeness order $\varepsilon$ .

As a conclusion of this study, we get uniform approximates of $x(t)$ and $y(t)$ on $[0, 1]$ . Indeed, first we determine $(\bar{X}, \bar{Y})$ as above, and get $(\bar{x}, \bar{y})$ , which are close to $(x, y)$ on $[t_2, 1]$ , with $\frac{1-t_2}{\varepsilon}$ infinitely large and $t_2 \sim 1$ .
Now on $[0, t_2]$ , we have $x(t) \sim w(t)$ and $y(t) \sim 0$ , whereas $\bar{x}(t) \sim \bar{B}$ , $\bar{y}(t) \sim 0$ (for $T$ infinitely large, $(\bar{X}(T), \bar{Y}(T))$ is close to the limit point) ; hence, as $w(t) \sim \bar{B}$ on $[t_2, 1]$ and $\varepsilon b(t, w(t)) \sim 0$ on $[0, 1]$ , we get $x(t) \sim \bar{x}(t) + w(t) - \bar{B}$ and $y(t) \sim \bar{y}(t)$ on $[0, 1]$ , where $\bar{x}(t) = X(\frac{1-t}{\varepsilon})$ and $\bar{y}(t) = Y(\frac{1-t}{\varepsilon})$ .
Notice that $\bar{x}(t) + w(t) - \bar{B}$ is precisely what the classical technic would find as a formal solution, using matched asymptotics ; what is typical in our treatment is that we get the approximation a posteriori, after a precise observation of an actual solution.

6) If we want more precisions about the order of the approximation, we must study the difference $z(t) = x(t) - \bar{x}(t) - w(t) + \bar{B}$ , which satisfies a differential equation of the same type as $x(t)$ . But stronger differentiability assumptions on $a$ and $b$ are needed. We don't persue in this way, although it is possible to do it.

However, as announced in the theorem above, it is easy to prove that for any $t \in [0, 1[$ , not $\sim 1$ , $\frac{x(t) - w(t)}{\varepsilon}$ and $\frac{\dot{x}(t) - \dot{w}(t)}{\varepsilon}$ are finite, without further assumptions.

Indeed, as $b$ is standard bounded and lipschitzian, for every finite point $p_0 = (t_0, x_0, b(t_0, x_0))$ of $S$ , there is a vertical cone $C(p_0) = C^+(p_0) \cup C^-(p_0)$ (upper and lower half-cones) with vertex $p_0$ and non infinitesimal angle $\theta(p_0)$ , such that $S \cap C(p_0) = \{p_0\}$ (see fig. 3).

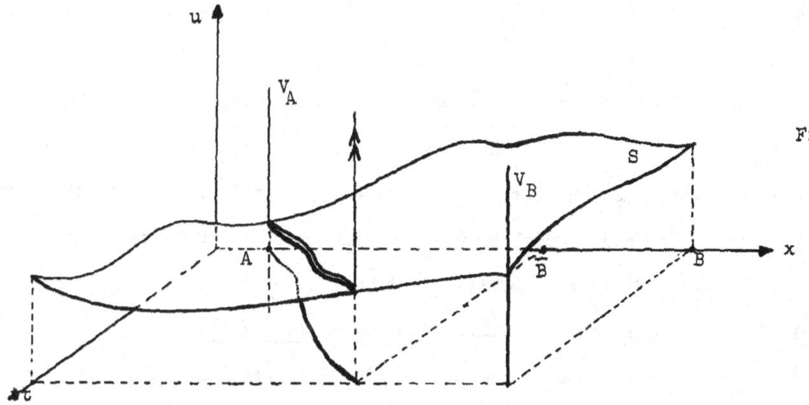


Fig.1

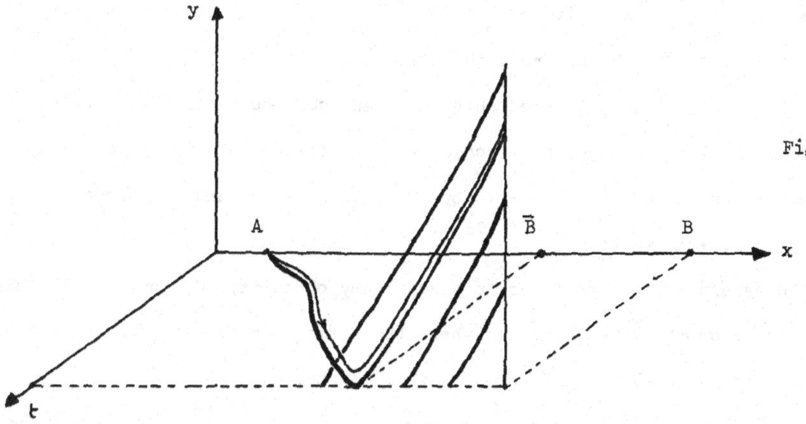


Fig.2

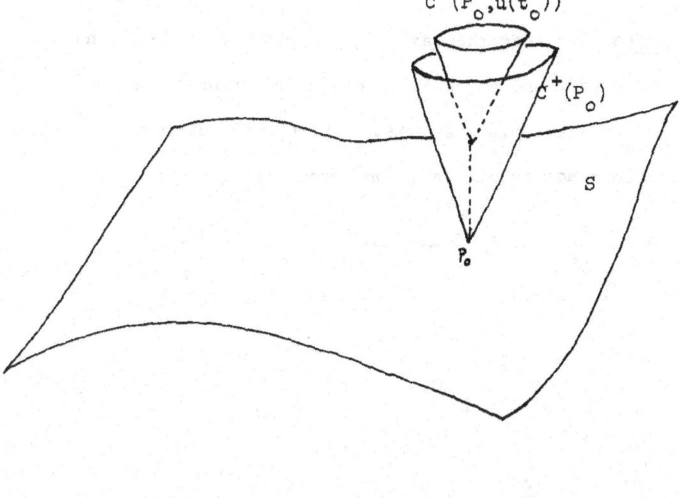


Fig.3

Now consider our solution $(t, x(t), u(t))$ of $(P_\varepsilon)$ and suppose that at some $t_0 \in [0,1]$, we have $\dot{u}(t_0) = \frac{u(t_0) - b(t_0, x(t_0))}{\varepsilon}$ infinitely large. Translate the half-cone $C^+(p_0)$ along the vertical from $p_0$ to $(t_0, x(t_0), u(t_0))$ (here $p_0 = (t_0, x(t_0), b(t_0, x(t_0)))$ ). You get a half cone $C^+(p_0, u(t_0))$, in which any point $(t, x, u)$ satisfies $u - b(t, x) > u(t_0) - b(t_0, x(t_0))$ ; hence in this cone, $\dot{u}(t)$ is infinitely large and quickly $u(t)$ becomes infinitely large, which is only possible if $t \sim 1$. Similar proof if $\dot{u}(t_0) < 0$.

Thus, as long as $t$ is not $\sim 1$, $\frac{u(t) - b(t, x(t))}{\varepsilon}$ is finite.

Use this improvement of lemma 2 to compute $z(t) = x(t) - w(t)$ as in § 3 and you get $\frac{x(t) - w(t)}{\varepsilon}$ finite for $t$ not $\sim 1$. Then $\frac{b(t, x(t)) - b(t, w(t))}{\varepsilon}$ is also finite and hence $\frac{\dot{x}(t) - \dot{w}(t)}{\varepsilon}$ is finite.

This ends the proof of the announced theorem.

Notice that until § 5, we only used uniqueness and continuity of the flows occuring in the proof. In § 6, we used the stronger assumption that $b$ is lipschitzian (this condition needs not be satisfied by $a$ ). In case $a$ and $b$ are of class $C^1$, all this clearly works.

We detailed carefully our proof, as to prevent any obscurity. Of course, all this is only an easy geometric observation where the permanence principle makes the join between the $(t, x, u)$-space and the $(t, x, y)$-space.

In the next lesson, we shall see a case where $a$ changes its sign with $x$. Free layers may occur and also solutions with different asymptotic behaviours.

7) Exercise. Consider the case $b = 0$, with $a$ only depending on $t$ : Compute $\bar{x}$ and prove that the approximate of $x$ is of order $\varepsilon$ on $[0, 1]$ (Hint : either study $x - \bar{x}$ by phase plane arguments, or use the maximum principle). In any case, you should not compute a second order term.

Lesson 12

A SEMI LINEAR PROBLEM WITH FREE AND BOUNDARY LAYERS

Problem. Describe the asymptotic behaviour, as $\varepsilon$ tends to $0$, of the solutions in the semi linear two-point boundary value problem :

$$P(a,\varepsilon)\begin{cases} \varepsilon\overset{''}{x} = x(a-\overset{'}{x}) & 0<t<1 \\ x(0)=A, \quad x(1)=B \text{ prescribed} \\ \varepsilon>0 \end{cases}$$

where $a$ is a real parameter.

THEOREM 1. <u>For any</u> $a\geq 0$, <u>and</u> $\varepsilon$ <u>little enough, problem</u> $P(a,\varepsilon)$ <u>has an unique solution with boundary or transition layers</u> (<u>for more precisions, see theorems 3 and 4 below</u>).

THEOREM 2. <u>For any</u> $a<0$, <u>and</u> $\varepsilon$ <u>little enough, problem</u> $P(a,\varepsilon)$ <u>has solutions whose number depends on the values of the boundary conditions. Moreover, there exist solutions with boundary and transition layers simultaneously</u> (<u>for more precisions, see theorem</u> 5 <u>below</u>).

Comments. 0) The choice of $[0,1]$ as time interval is not a restriction. If it is replaced by $[\alpha,\beta]$ by means of the change of variable $T=(t-\alpha)(\beta-\alpha)^{-1}$, you get problem $P(a(\beta-\alpha),\varepsilon(\beta-\alpha)^{-1})$. Moreover, substitute $x$ for $\frac{x}{|a|}$ (if $a\neq 0$), the problem reduces to the three particular cases $a=0$, $a=1$ and $a=-1$.

Problem $P(1,\varepsilon)$ has often been considered as a model quasi linear example for singular perturbations (see [Coc],[Col],[Ha],[Ho],[DPS]) provided its wide variety of interesting behaviours ; indeed, one or more solutions of the reduced equation $u(1-\overset{'}{u})=0$ may be required to approximate the solution, and the transition may occur near interior points of $[0,1]$.

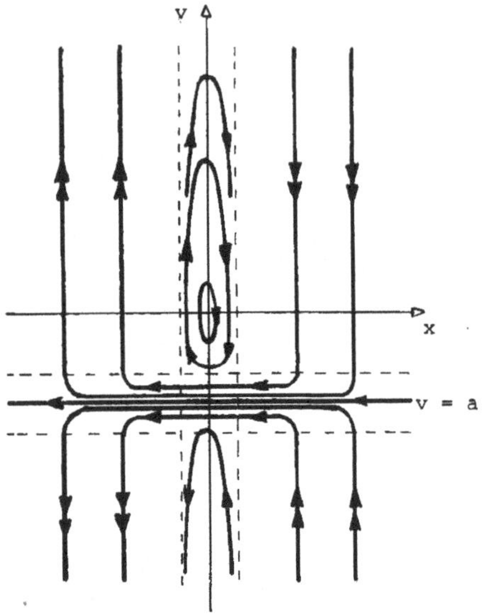


FIG. 1i a < 0

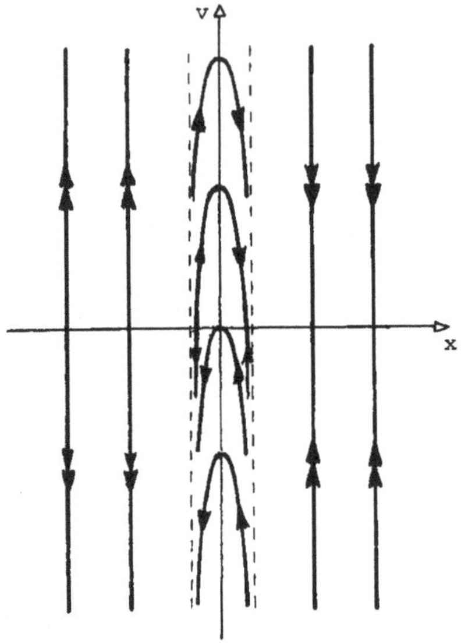


FIG. 1ii a = 0

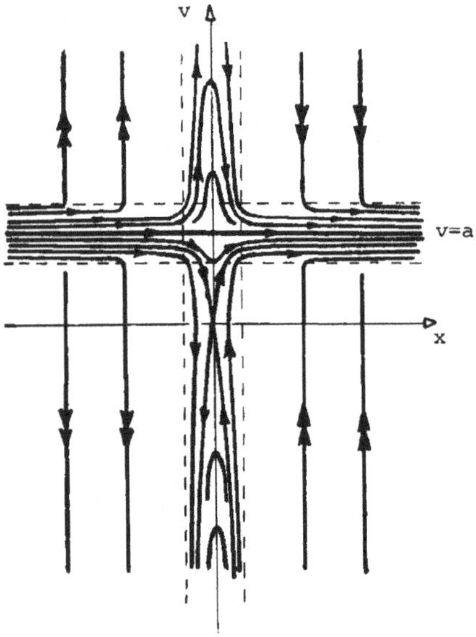


FIG. 1iii a > 0

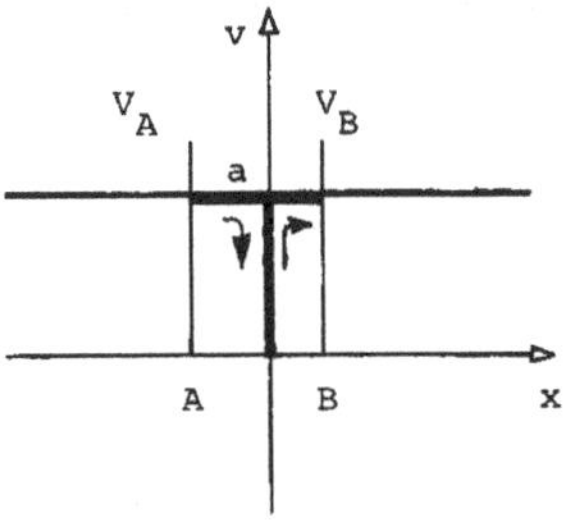


FIG. 2. The shadow of the integral curve solving $P(a\varepsilon)$ in case $a > 0$ and $0 \leq B < A+a \leq a$

Problem $P(-1, \varepsilon)$ is related to the theory of tubulence ; if we replace $x$ by $-x$ in $P(-1, \varepsilon)$, we get the differential equation $\varepsilon\ddot{x} = x\dot{x} - x$, which is the steady state version of Burgers equation (see [Bu], [Ho], [M]).

1) In the non-standard treatment, we assume, as usually, that $a$, $A$ and $B$ are standard, and we fix $\varepsilon \sim 0$. Then, the vector field $X_1(a, \varepsilon)$ of components $(v, \frac{x}{\varepsilon}(a-v))$ in the phase plane $(x, v=\dot{x})$, associated with the differential equation $\varepsilon\ddot{x} = x(a-\dot{x})$ has a slow-fast flow. Following lesson IV.5, its integral curves in the finite plane are nearly vertical, excepted when $x$ or $a-v$ is of order $\varepsilon$. If $a=0$, then all the points of the x-axis are singular points. If $a \neq 0$, then $v=a$ is a particular solution, and the point $(0, 0)$ is an isolated singular point (a center if $a<0$, a saddle point if $a>0$). Using the reflexion of the integral curves in the v-axis, we get the phase plane portrait described by figure 1. To solve problem $P(a, \varepsilon)$, we have to find some integral curve, starting on $V_A$ (the vertical of $A$) and reaching $V_B$ at time $1$. We see that, excepted when $B = A + a$ (where $x(t) = at + A$ is the solution of $P(a, \varepsilon)$), only in one case, there exist such an integral curve in the finite plane, namely whenever $a>0$ and $0 \leq B < A + a \leq a$ (its shadow is represented in figure 2 ; see also theorem 4, case iv.). Indeed, in all other cases, any integral curve, starting on $V_A$, either reaches $V_B$ before time $1$ or after time $1$, or leaves the finite plane. Thus we need informations for the infinitely large values of $v$.

2) Let us observe the integral curves in the Lienard plane $(x, u = \varepsilon\dot{x} + \frac{x^2}{2})$ ; we get the vector field $X_2(a, \varepsilon)$ of components $(\frac{1}{\varepsilon}(u - \frac{x^2}{2}), ax)$ which has its first component infinitely large ; its integral curves are nearly horizontal outside the halo of the slow curve $u = \frac{x^2}{2}$ (more precisely for any point with $\frac{2u - x^2}{\varepsilon}$ infinitely large). As $u = \varepsilon a + \frac{x^2}{2}$ is a particular solution, using the reflexions in the u-axis, we get the shadows of the integral curves as shown in figure 3.

Then, for any integral curve, we have two approximations : the first (given by figure 1) asserts that, for finite values of $v$, the solution is $\varepsilon$-close to

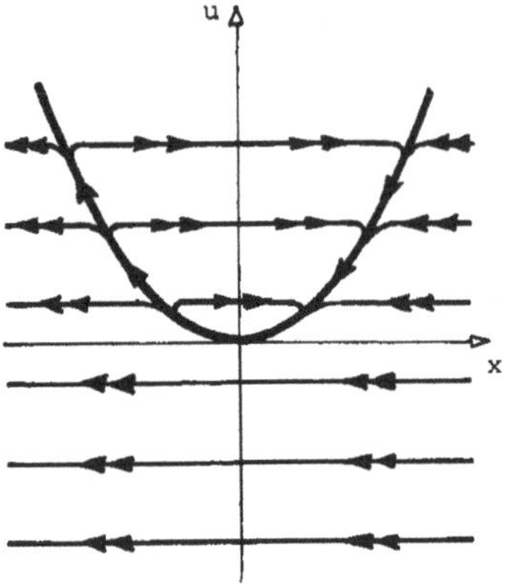


FIG. 3i. $a < 0$

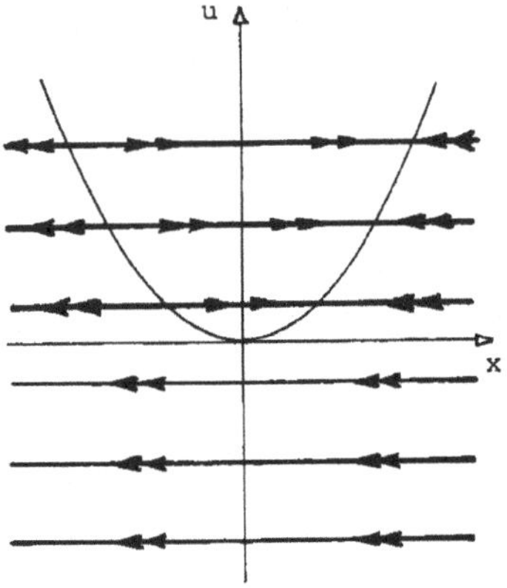


FIG. 3ii. $a=0$

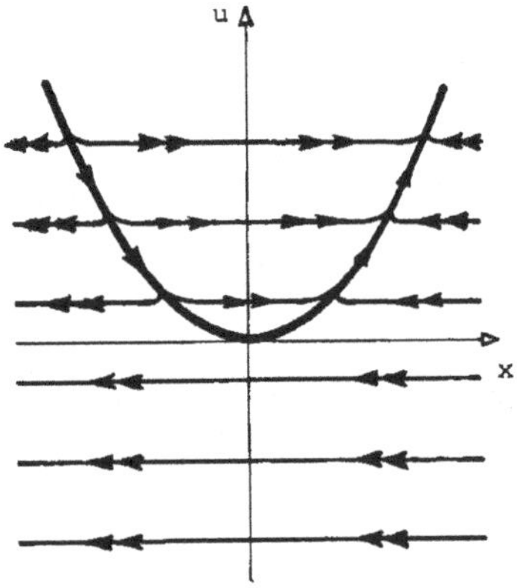


FIG. 3iii. $a > 0$

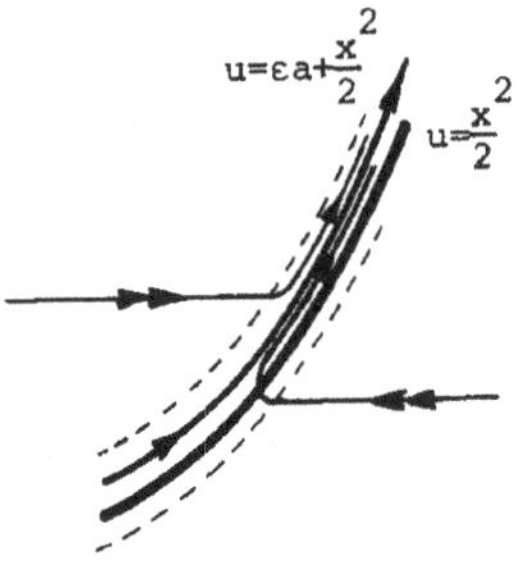


FIG . 3iv.

FIG. 3. The shadows of the integral curves in the Lienard plane $(x,u)$. Fig. 3iv shows what happens at a corner (in case $a > 0$).

$v = a$ or $x = 0$ (i.e. the solutions of the reduced equation) nearly in the whole of $[0, 1]$, and quick jumps must arise ; the second one (given by figure 3) asserts that during these quick jumps (i.e. for infinitely large values of $v$), the integral curve is such that $\varepsilon \dot{x}(t) + \frac{x^2(t)}{2}$ is nearly constant. To match both behaviour together, we use as usual the permanence principle. Indeed, observe an integral curve in both $(x, u)$ and $(x, v)$ planes : for any finite $v$ (not $\varepsilon$-close to $a$), $x(t)$ is infinitely close to some standard value $\bar{x}$ (see figure 4.i.) : this property is permanent until some infinitely large value of $v$, for which the second behaviour occurs ; hence, in the $(x, u)$ plane, the curve reaches the halo of the parabola $2u = x^2$ near the point $\bar{x}$ (see figure 4.ii).

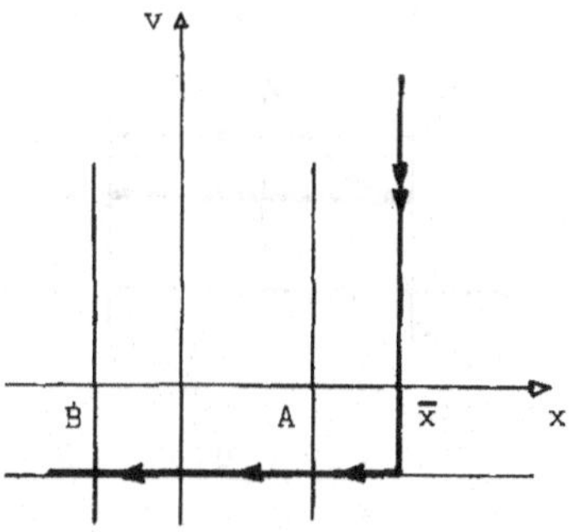


Fig. 4i

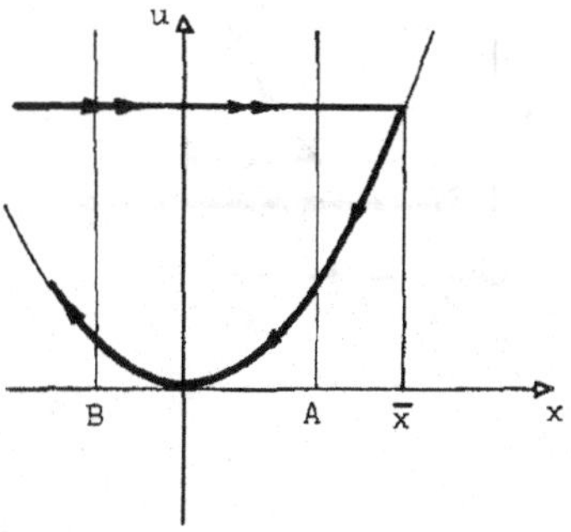


Fig.4ii

3) Let us outline the solution of problem $P(a, \varepsilon)$ in the particular case $0 < A < B - a$. Put $\bar{x} = B - a$, then starting on $V_A$, the integral curve of figure 4 reaches $V_{\bar{x}}$ after an infinitesimal time ; after this, the speed is nearly $a$, and hence at time $1$, the moving point is nearly on $V_B$. Thus we may expect a solution of problem $P(a, \varepsilon)$ with a boundary layer at $t = 0$. Before going on, let us make a remark on the thickness of this layer. After any time $t$ with $t/\varepsilon$ infinitely large, the moving point is on the right of $V_{\bar{x}}$ ; moreover, after a time with $t/\sqrt{\varepsilon}$ infinitely large, it reaches a region where $\dot{x}$ is

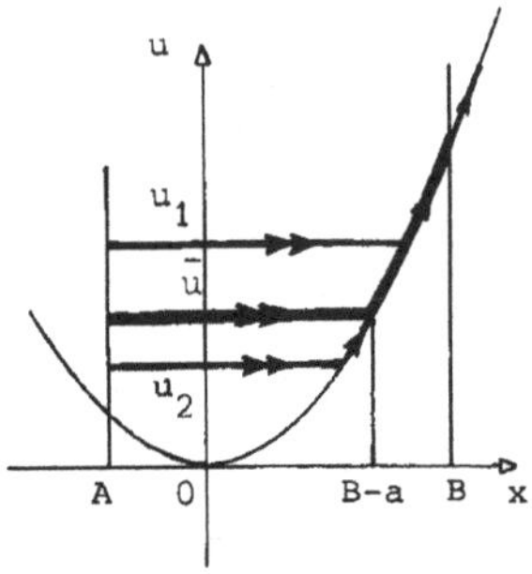


a-B≤A<B-a

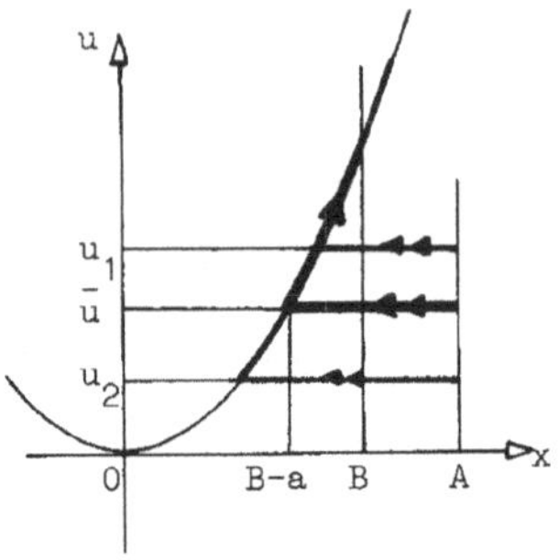


0≤B-a<A

FIG. 5.

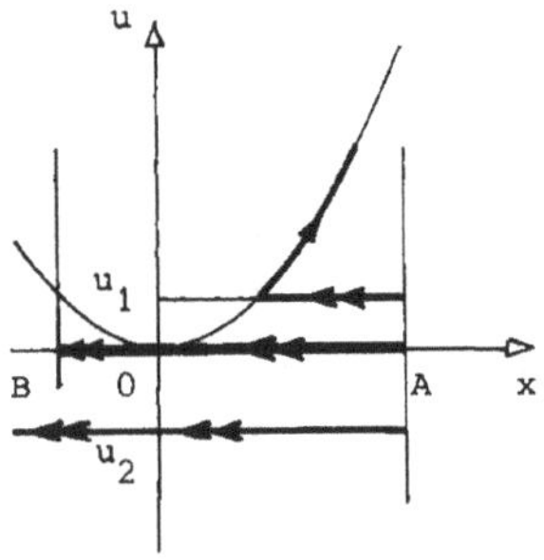


B<0<A

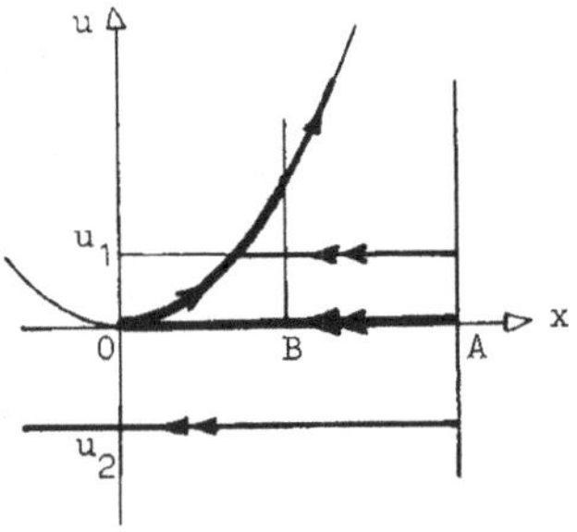


0≤B<a, 0<A

FIG.6.

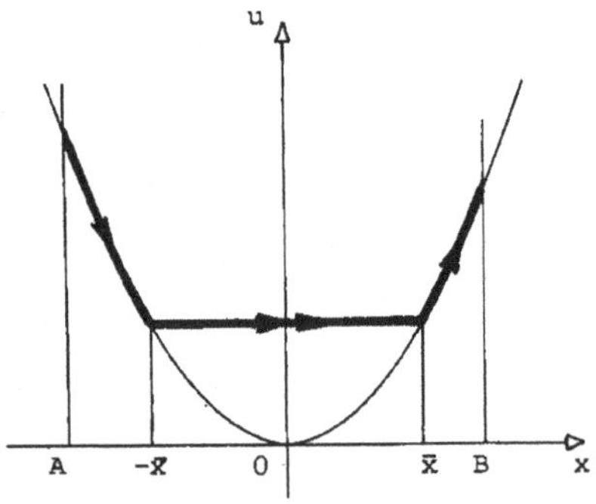


FIG. 7.

$\varepsilon$-close to $a$. Thus, the layers of $x(t)$ (resp. $x'(t)$ ) are of thickness order $\varepsilon$ (resp. $\sqrt{\varepsilon}$ ).

4) Proof of theorem 1. Assume $a>0$ (the case $a=0$ is easier and may be handled in the same manner). Let $x(t;x_o,u_o)$ be the first component of the integral curve of the vector field $X_2(a,\varepsilon)$, starting at $(x_o,u_o)$. We restrict our attention to the three representative cases :

i) $A+B \geq a$ and $B>a$. Then $x(1;A,u_1)>B$ for every standard $u_1>\bar{u}$, where $\bar{u}=\frac{(B-a)^2}{2}$, and $x(1;A,u_2)<B$ for every standard $u_2<\bar{u}$ (see figure 5) ; due to the continuity of $x(1;A,u_o)$ with respect to $u_o$, there exists some value of $u_o$ (infinitely close to $\bar{u}$ ) such that $x(1;A,u_o)=B$. The corresponding solution $x(t)=x(t;A,u_o)$ of $P(a,\varepsilon)$ has boundary layer character at time $0$ and its shadow is

$$°x(t) = \begin{cases} A & \text{at } t=0 \\ at+B-a & \text{on } (0,1]. \end{cases}$$

We say that $x(t)$ is approximated by $at+B-a$ on $(0,1]$, (precisely on every $[\delta,1]$ with $\delta$ standard positive).

ii) $A>0$ and $B\leq a$. Then $x(1;A,u_1)>a\geq B$ for every standard $u_1>0$, and for every standard $u_2<0$, $x(1;A,u_2)$ is infinitely large negative (see figure 6). Then for some infinitesimal $u_o$, $x(t;A,u_o)$ is the solution of $P(a,\varepsilon)$. If $B\geq 0$, this solution has boundary layer character at time $0$ and its shadow is

$$°x(t) = \begin{cases} 0 & \text{on } (0,1-B/a] \\ at+B-a & \text{on } [1-B/a,1]. \end{cases}$$

If $B<0$, the solution exhibits boundary layers at $t=0$ and $t=1$, and is approximated by $0$ on $(0,1)$, i.e. its shadow verify $°x(0)=A$, $°x(1)=B$ and $°x(t)=0$ on $(0,1)$.

iii) We look now for a solution with a free layer. Such a solution must start at some $A<0$, have a jump from $-\bar{x}$ to $\bar{x}$, and then go from $\bar{x}$ to $B$ (see figure 7). From the conditions $B-\bar{x}-\bar{x}-A=a$, $\bar{x}>0$, $B-\bar{x}>0$ and

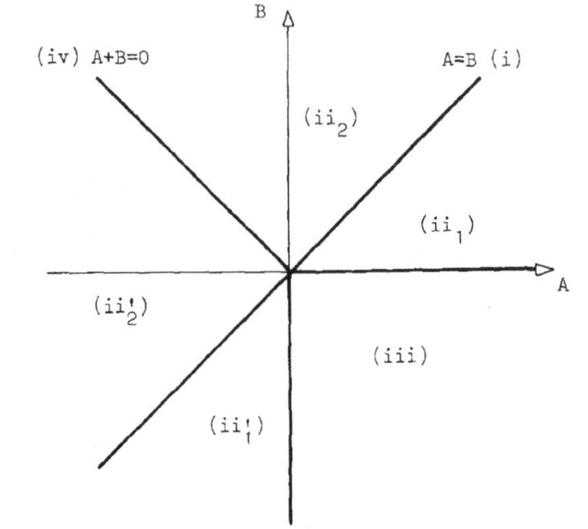


FIG. 8i. a = 0

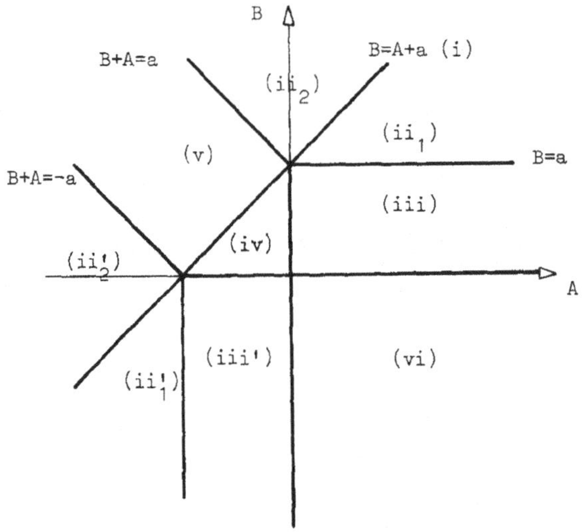


FIG. 8ii. $a > 0$

Boundary value portrait of $P(a, \varepsilon)$ when $a \geq 0$.

The corresponding solutions are described in theorem 3, fig. 9 ($a=0$) and theorem 4, fig.10 ($a>0$).

$A+\bar{x}<0$ , we infer $B-A>a$ , $|A+B|<a$ and $\bar{x}=\frac{B-A-a}{2}$ . We may prove the existence of the solution as above. Its shadow is

$$°x(t) = \begin{cases} at+A & \text{on } [0, \frac{1}{2}-\frac{A+B}{2a}) \\ at+B-a & \text{on } (\frac{1}{2}-\frac{A+B}{2a}, 1] . \end{cases}$$

Thus, we may state more precisely theorem 1. We get

THEOREM 3. Let $x(t)$ be the solution of $P(0,\varepsilon)$ and $°x(t)$ its shadow :

i) if $A=B$ , then $x(t)=A$ on $[0,1]$ ;

ii) if $A+B>0$ , $B\geq 0$ and $A\neq B$ (resp. $A+B<0$ , $A\leq 0$ and $A\neq B$ ), then :

$$°x(t) = \begin{cases} A & \text{at } t=0 \\ B & \text{on } (0,1] \end{cases} \quad (\text{resp. } °x(t) = \begin{cases} A & \text{on } [0,1) \\ B & \text{at } t=1 \end{cases}).$$

There is a boundary layer of thickness order $\varepsilon$ at $t=0$ (resp. $t=1$ ).

iii) if $A>0$ and $B<0$ , then $°x(0)=A$ , $°x(1)=B$ and $°x(t)=0$ on $(0,1)$ . There is boundary layers of thickness order $\varepsilon$ at $t=0$ and $t=1$ .

iv) if $A+B=0$ and $B>0$ , then

$$°x(t) = \begin{cases} A & \text{on } [0,\frac{1}{2}) \\ B & \text{on } (\frac{1}{2},1] . \end{cases}$$

There is a transition layer of thickness order $\varepsilon$ at $t=\frac{1}{2}$ .

We summarize these results in the plane $(A,B)$ (see figure 8.i) and represent the geometric shadows of the solutions in figure 9.

THEOREM 4. Let $x(t)$ be the solution of $P(a,\varepsilon)$ , $a>0$ and $°x(t)$ its shadow.

i) if $B=A+a$ , then $x(t) = at+A$ on $[0,1]$ .

ii) if $B\geq a$ , $A+B\geq a$ and $B\neq A+a$ (resp. $A+B\leq -a$ , $A\leq -a$ and $B\neq A+a$ ), then :

$$°x(t) = \begin{cases} A & \text{at } t=0 \\ at+B-a & \text{on } (0,1] \end{cases} \quad (\text{resp. } °x(t) = \begin{cases} at+A & \text{on } [0,1) \\ B & \text{at } t=1 \end{cases}).$$

There is a boundary layer of thickness order $\varepsilon$ at $t=0$ (resp. $t=1$ ).

iii) if $0\leq B<a$ and $A>0$ (resp. $-a<A\leq 0$ and $B<0$ ), then

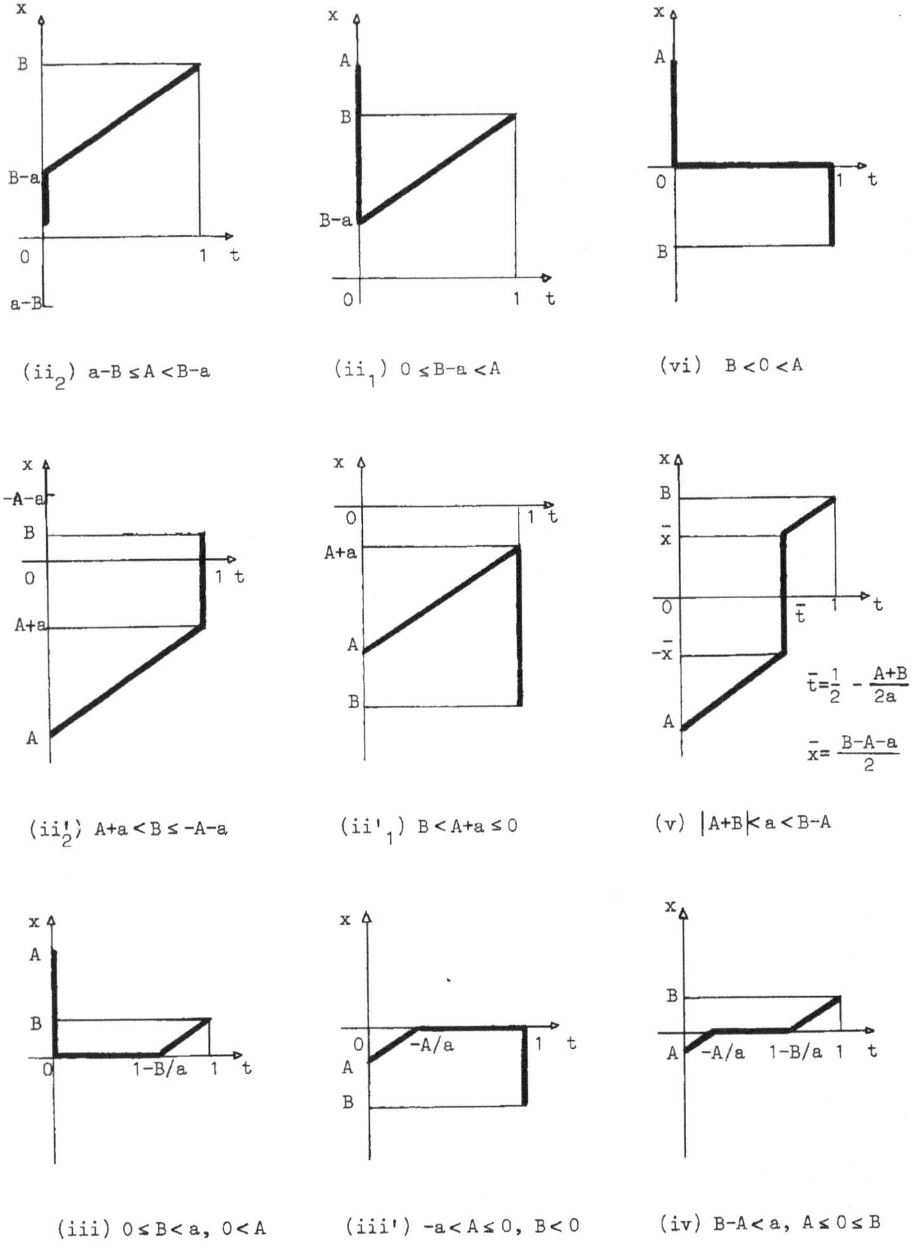


FIG. 10. The geometric shadows of the solutions of $P(a,\varepsilon)$, $a > 0$.

$$°x(t) = \begin{cases} A \ \underline{\text{at}} \ t=0 \\ 0 \ \underline{\text{on}} \ (0, 1-B/a] \\ at+B-a \ \underline{\text{on}} \ [1-\frac{B}{a}, 1] \end{cases} \qquad (\underline{\text{resp.}} \ °x(t) = \begin{cases} at+A \ \underline{\text{on}} \ [0, -\frac{A}{a}] \\ 0 \ \underline{\text{on}} \ [-\frac{A}{a}, 1) \\ B \ \underline{\text{at}} \ t=1 \end{cases} ).$$

<u>There is a boundary layer of thickness order</u> $\varepsilon$ <u>at</u> $t=0$ (<u>resp.</u> $t=1$ ) <u>and</u> $\dot{x}(t)$ <u>has a transition layer of thickness order</u> $\sqrt{\varepsilon}$ <u>at</u> $t = 1-B/a$ (<u>resp.</u> $t=-A/a$ ).

iv) <u>if</u> $B-A<a$ , $A \leq 0$ <u>and</u> $B \geq 0$ , <u>then</u>

$$°x(t) = \begin{cases} at+A \ \underline{\text{on}} \ [0, -A/a] \\ 0 \ \underline{\text{on}} \ [-A/a, 1-B/a] \\ at+B-a \ \underline{\text{on}} \ [1-B/a, 1] . \end{cases}$$

<u>And</u> $\dot{x}(t)$ <u>has transition layers of thickness order</u> $\sqrt{\varepsilon}$ <u>at</u> $t=-A/a$ <u>and</u> $t=1-B/a$ .

v) <u>if</u> $|A+B| < a$ <u>and</u> $B-A>a$ , <u>then</u>

$$°x(t) = \begin{cases} at+A \ \underline{\text{on}} \ [0, \frac{1}{2}-\frac{A+B}{2a}) \\ at+B-a \ \underline{\text{on}} \ (\frac{1}{2}-\frac{A+B}{2a}, 1] \end{cases}$$

<u>There is a transition layer of thickness order</u> $\varepsilon$ <u>at</u> $t = \frac{1}{2} - \frac{A+B}{2a}$ .

We summarize these results in the plane $(A, B)$ (see figure 8.ii) and represent the geometric shadows of the solutions in figure 10.

5) Note that we get all these results, only by means of phase plane arguments (compare with the classical treatment in the papers listed in section 0). The case $a<0$ is even more interesting ; at our knowledge, the classical litterature didn't describe the solutions for all choices of the boundary conditions (see [Ho], p. 84 or [M]). Our aim is to give a complete description of the solutions of $P(a, \varepsilon)$ with $a<0$ . Due to the occurence of periodic solutions, we may obtain more than one solution, by spinning around these periodic solutions.

6) <u>Proof of theorem 2</u>. First assume that $B<A+a$ . Then, the velocity $\dot{x}$ must be taken lower than $a$ , and we are not in the region of periodic solutions. Three cases have to be considered.

i) $2A>-a$ and $2B<a$ . Then $x(1; A, u_1) > \frac{a}{2} > B$ for every standard

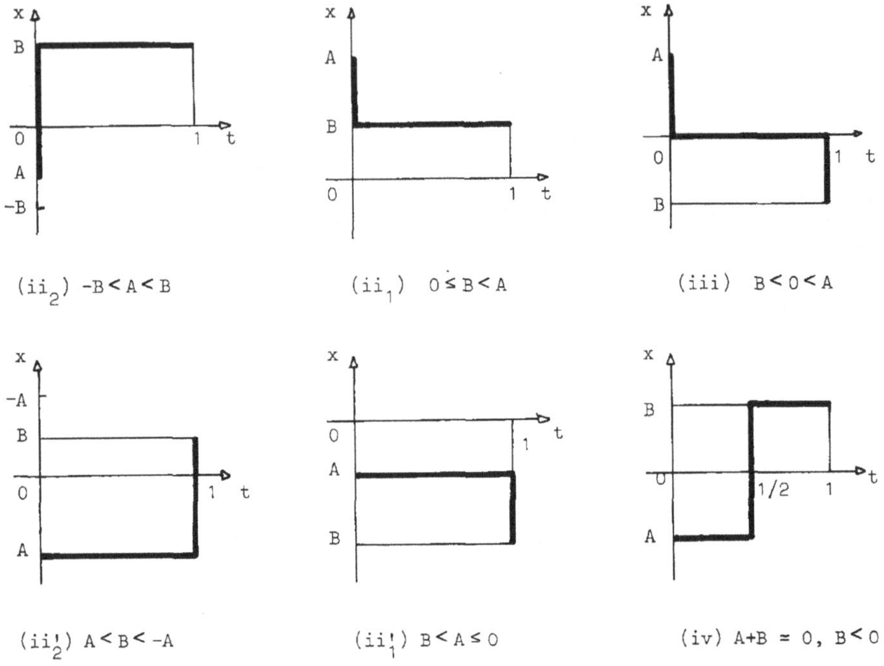


FIG.9. Geometric shadows of the solutions of $P(0,\epsilon)$.

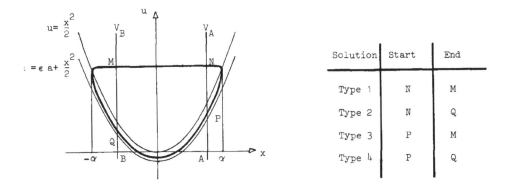

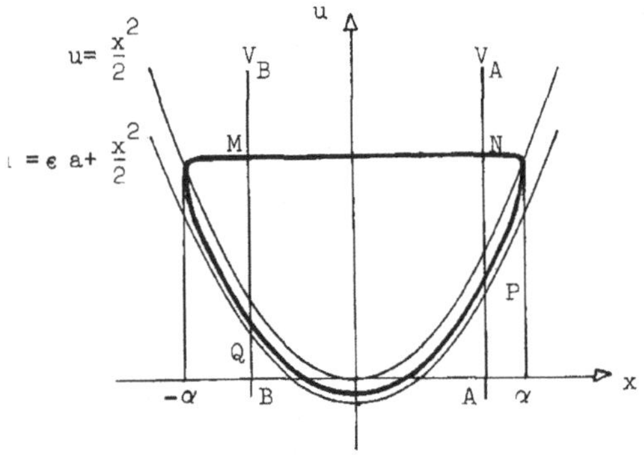


| Solution | Start | End |
|---|---|---|
| Type 1 | N | M |
| Type 2 | N | Q |
| Type 3 | P | M |
| Type 4 | P | Q |

FIG. 13. The four types of solutions of $P(a,\epsilon)$, when $a<0$ and $B>A+a$.

$u_1 > \bar{u}$ , where $\bar{u} = \frac{a^2}{8}$ and for every standard $u_2 < \bar{u}$ , $x(1 ; A , u_2)$ is infinitely large negative (see figure 11.). It then follows that there must be some value $u_o$ (infinitely close to $\bar{u}$ ) such that $x(1 ; A , u_o) = B$ . The solution $x(t ; A , u_o)$ of $P(a , \varepsilon)$ , with boundary layers at $t = 0$ and $t = 1$ is approximated by $at - \frac{a}{2}$ on $(0 , 1)$ (see figure 12.i).

ii) $2B \geq a$ and $A > B - a$ . Then $x(1 ; A , u_1) > B$ for every standard $u_1 > \bar{u}$ , where $\bar{u} = \frac{(B-a)^2}{2}$ , and $x(1 ; A , u_2) < B$ for every standard $u_2 > \bar{u}$ (see figure 11.i). Then, for some $u_o$ (infinitely close to $\bar{u}$ ), $x(t ; A , u_o)$ is the solution of problem $P(a , \varepsilon)$ . It has a boundary layer character at $t = 0$ and is approximated by $at + B - a$ on $(0 , 1]$ (see figure 12.ii).

iii) $2A \leq -a$ and $B < A + a$ . As previously, we may prove that the solution exhibits a boundary layer at $t = 1$ and is approximated by $at + A$ on $[0 , 1)$ (see figures 11.iii and 12.iii).

7) Proof of theorem 2 (continued). When $B > A + a$ , we must take $x'$ larger than $a$ and then all the integral curves are periodic orbits of the vector field $X_2(a , \varepsilon)$ . We have four types of solutions (see figure 13). Necessary conditions of their existence are $|A| \leq \alpha$ and $|B| \leq \alpha$ ; we call them conditions (*), and we tell $x_i^k$ $(i = 1 , 2 , 3 , 4 ; \ k = 0 , 1 , 2 , \ldots)$ the solutions of type $i$ which spins $k$ times around the center $(0 , 0)$ :

i) Type 1. We have $2k\alpha + 2\alpha = -a$, then $\alpha = -a/2(k+1)$ , and from conditions (*) we infer $|A| \leq -a/2(k+1)$ and $B \leq -a/2(k+1)$ . Let $t_h = \frac{h}{k+1}$, $h = 0 , \ldots , k+1$ . The solution $x_1^k$ has boundary layers at $t_o = 0$ and $t_{k+1} = 1$ , and $k$ transitions layers at $t_h$ , $h = 1 , \ldots , k$ . Its shadow is

$$°x_1^h(t) = \begin{cases} A & \text{at } t = 0 \\ at + (2h+1)\alpha & \text{on } (t_h , t_{h+1}) \qquad h = 0 , \ldots , k . \\ B & \text{at } t = 1 \end{cases}$$

ii) Type 2. We have $2k\alpha + \alpha - B = -a$ , then $\alpha = (B-a)/(2k+1)$ , and from conditions (*), we infer $|A| \leq (B-a)/(2k+1)$ and $a/2(k+1) \leq B \leq -a/2k$ (when $k = 0$ , replace this condition by $B \geq a/2$ ). Let $t_h = \frac{2h(a-B)}{(2k+1)a}$ , $h = 0 , \ldots , k$ . The solution $x_2^k$ has a boundary layer at $t_o = 0$ and $k$ transitions layers at

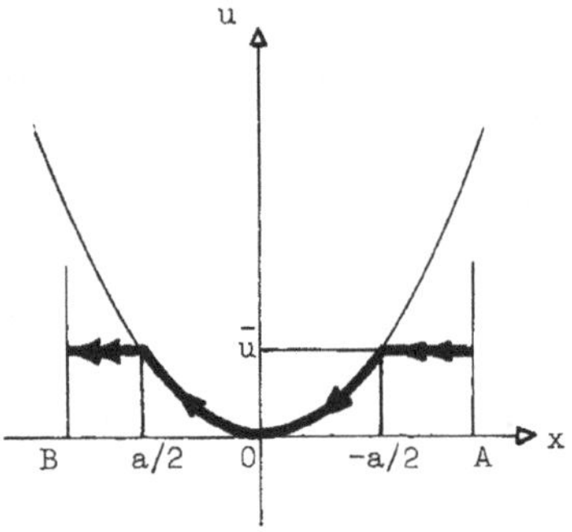


FIG. 11i. $A > -a/2$, $B < a/2$

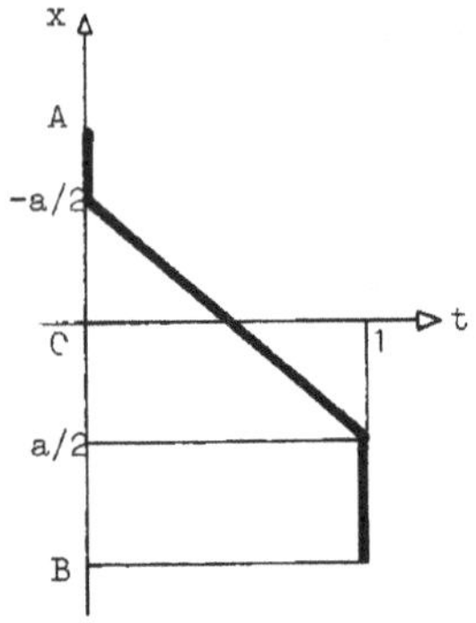


FIG. 12i.

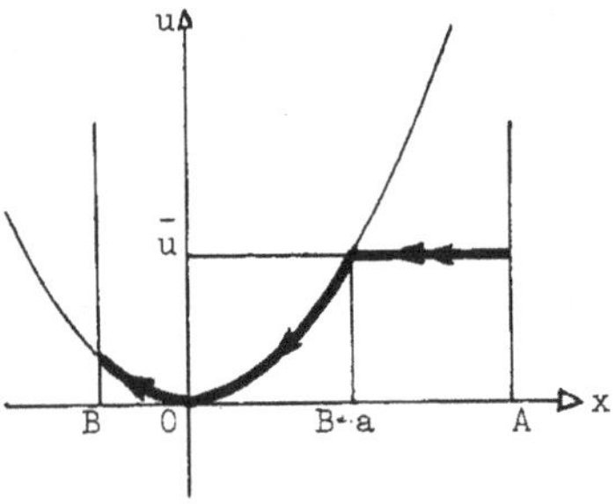


FIG. 11ii. $B \geq a/2$, $a+A > B$

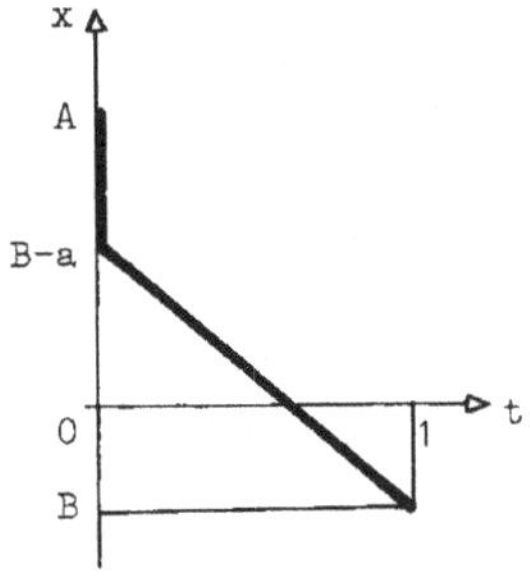


FIG. 12ii.

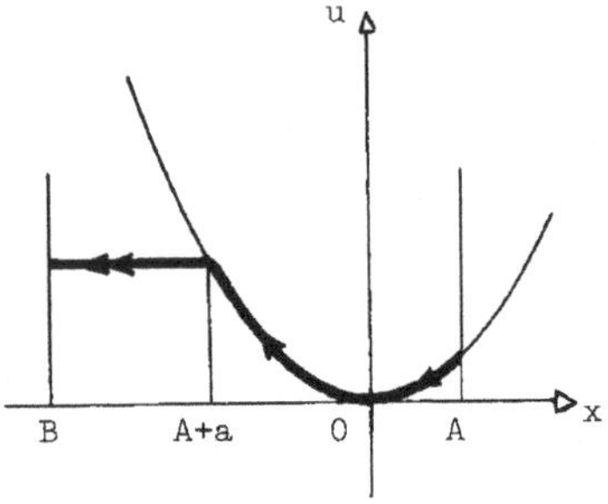


FIG. 11iii. $A \leq -a/2$, $A+a > B$

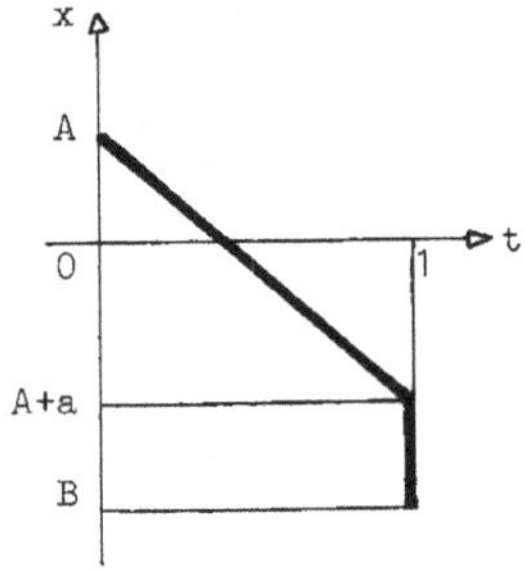


FIG. 12iii.

The solutions of $P(a, \epsilon)$ when $a < 0$ and $B < A+a$

$t_h$ , $h=1,\dots,k$ . Its shadow is

$$°x_2^k(t) = \begin{cases} A & \text{at} \quad t=0 \\ at+(2h+1)\alpha & \text{on} \quad (t_h\,,\ t_{h+1}) \qquad h=0,\dots,k-1 \\ at+B-a & \text{on} \quad (t_k\,,1] \end{cases}$$

iii) Type 3. We have $2k\alpha + A + \alpha = -a$, then $\alpha = -(A+a)/(2k+1)$ , and from condition (*) we infer $a/2k \le A \le -a/2(k+1)$ (when $k=0$ , replace this condition by $A \le -a/2$) and $|B| \le -(A+a)/(2k+1)$ . Let $t_h = \dfrac{(2h+1)a-2(k-h)A}{(2k+1)a}$ , $h=0,\dots,k$ . The solution $x_3^k$ has a boundary layer at $t_k=1$ , and $k$ transitions layers at $t_h$ , $h=0,\dots,k-1$ . Its shadow is

$$°x_3^k(t) = \begin{cases} at+A & \text{on} \quad [0\,,t_o) \\ at+A+2h\alpha & \text{on} \quad (t_h\,,t_{h+1})\,, \quad h=0,\dots,k-1\,. \\ B & \text{at} \quad t=1 \end{cases}$$

iv) Type 4. We have $2k\alpha + A + \alpha + \alpha - B = -a$ , then $\alpha = (B-A-a)/2(k+1)$ , and from conditions (*) we infer

$(a-B)/(2k+1) \le A \le (B-a)/(2k+3)$ and $(A+a)/(2k+3) \le B \le -(A+a)/(2k+1)$ .

Let $t_h = \dfrac{(2h+1)(a-B)-(2k-2h+1)A}{2(k+1)a}$ , $h=0,\dots,k$ . The solution $x_4^k$ has $k+1$ transitions layers at $t_h$ , $h=0,\dots,k$ , its shadow is

$$°x_4^k(t) = \begin{cases} at+A & \text{on} \quad [0\,,t_o) \\ at+A+2h\alpha & \text{on} \quad (t_h\,,t_{h+1}) \qquad h=0,\dots,k-1 \\ at+B-a & \text{on} \quad (t_k\,,1]\,. \end{cases}$$

In figure 14, we represent the geometrical shadows of these solutions, and their domains of existence in the plane $(A\,,B)$ .

8) Proof of theorem 2 (end). All the solutions $x_i^k$ are different, excepted when the point $(A\,,B)$ lies on the boundaries of their domains of existence. Let us examine this case ; for any $k>0$ :

i) if $(A\,,B) \in (M_{k+1}\,,N_{k+1})$ (resp. $(A\,,B) \in (Q_{k+1}\,,P_{k+1})$ ), then $x_1^k = x_3^k$ (resp. $x_1^k = x_3^{k+1}$ ).

ii) if $(A\,,B) \in (P_{k+1}\,,N_{k+1})$ (resp. $(A\,,B) \in (Q_{k+1}\,,M_{k+1})$ ), then $x_1^k = x_2^k$ (resp. $x_1^k = x_2^{k+1}$ ).

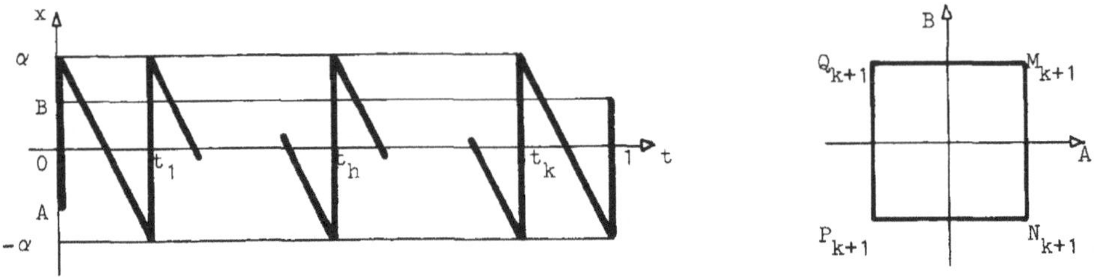


(i) i=1 $\alpha = \frac{-a}{2(k+1)}$ $t_h = \frac{h}{k+1}$

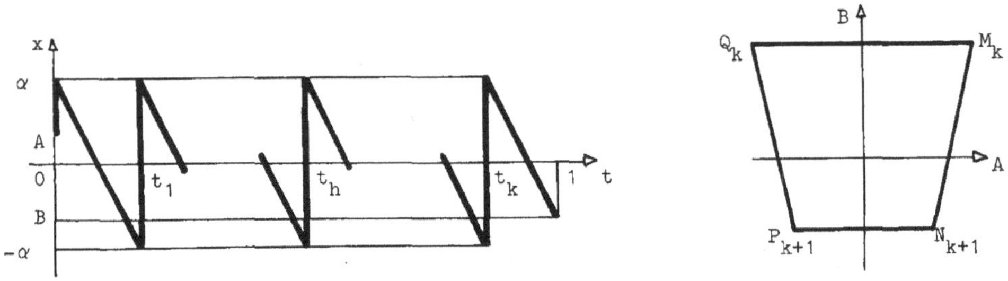


(ii) i=2 $\alpha = \frac{B-a}{2k+1}$ $t_h = \frac{2h(a-B)}{(2k+1)a}$

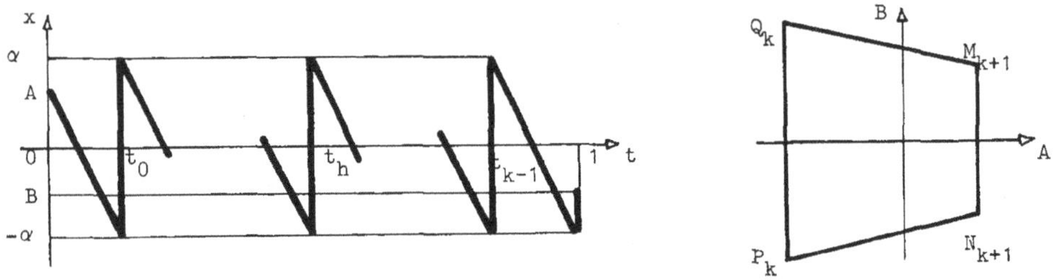


(iii) i=3 $\alpha = -\frac{A+a}{2k+1}$ $t_h = \frac{(2h+1)a - 2(k-h)A}{(2k+1)a}$

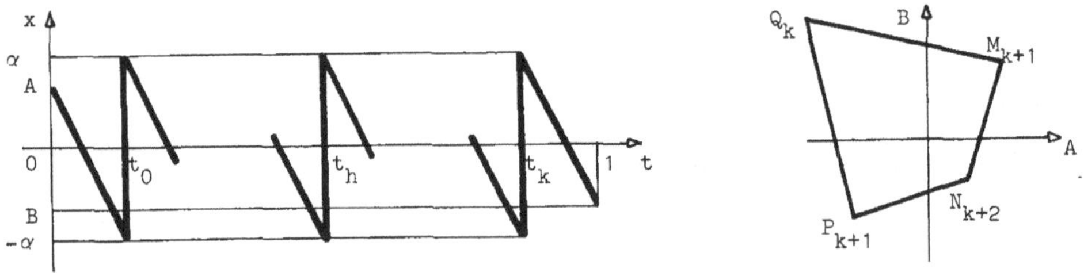


(iv) i=4 $\alpha = \frac{B-A-a}{2(k+1)}$ $t_h = \frac{(2h+1)(a-B) - (2k-2h+1)A}{2(k+1)a}$

FIG.14. The solutions of $P(a,\varepsilon)$ when $a < 0$ and $B > A+a$.

iii) if $(A, B) \in (Q_k, P_{k+1})$ (resp. $(A, B) \in (M_k, N_{k+1})$ ), then $x_2^k = x_4^k$ (resp. $x_2^k = x_4^{k-1}$ ).

iv) if $(A, B) \in (Q_k, M_{k+1})$ (resp. $(A, B) \in (P_k, N_{k+1})$ ), then $x_3^k = x_4^k$ (resp. $x_3^k = x_4^{k-1}$ ).

v) if $(A, B) = M_{k+1}$ (resp. $(A, B) = P_{k+1}$ ), then $x_1^k = x_2^{k+1} = x_3^k = x_4^k$ (resp. $x_1^k = x_2^k = x_3^{k+1} = x_4^k$ ).

vi) if $(A, B) = N_{k+1}$ (resp. $(A, B) = Q_{k+1}$), then $x_1^k = x_2^k = x_2^k = x_4^{k-1}$ (resp. $x_1^k = x_2^{k+1} = x_3^{k+1} = x_4^{k+1}$ ).

We have then proved the following theorem :

THEOREM 5. Let $x(t)$ be the solution of $P(a, \varepsilon)$, $a<0$, and $°x(t)$ its shadow (see figure 15) :

i) if $B = A + a$, then $x(t) = at + A$ on $[0, 1]$.

ii) if $(A, B) \in I_o = \{(A, B) ; 2A \geq -a, 2B \leq a, (A, B) \neq (-a/2, a/2)\}$, then $°x(0) = A$, $°x(1) = B$ and $°x(t) = at - \frac{a}{2}$ on $(0, 1)$ (see figure 12.i).

iii) if $(A, B) \in II_o$ (resp. $II'_o$ ), then

$$°x(t) = \begin{cases} A & \text{at } t = 0 \\ at + B - a & \text{on } (0, 1] \end{cases} \qquad \left(\text{resp. } °x(t) = \begin{cases} at + A & \text{on } [0, 1) \\ B & \text{at } t = 1 \end{cases}\right).$$

In case $II_o(1)$ (resp. $II'_o(1)$ ), see figure 12.ii (resp. 12.iii).

In case $II_o(2)$ (resp. $II'_o(2)$ ), see figure 14.ii (resp. 14.iii) for $k = 0$.

iv) if $(A, B) \in III_o$, then there exists three solutions which are $x_2^o$, $x_3^o$ and $x_4^o$ (see figure 14).

v) if $(A, B) \in \{|A + B| = -a, B \geq A\} \cup \{2A = -a, |2B| < -a\} \cup \{2B = a, |2A| < -a\}$, there exists two solutions which are $x_2^o$ and $x_3^o$ (see figure 14).

And for every standard $n \geq 1$ :

vi) if $(A, B) \in I_n$, there exists $4n - 1$ solutions which are $x_1^k$, $x_2^k$ and $x_3^k$ for $k = 0, \ldots, n-1$ and $x_4^k$ for $k = 0, \ldots, n-2$.

vii) if $(A, B) \in II_n$ (resp. $II'_n$ ), there exist $4n + 1$ solutions, which are $x_1^k$, $x_2^k$, $x_3^k$, $x_4^k$ for $k = 0, \ldots, n-1$ and $x_2^n$ (resp. $x_3^n$ ).

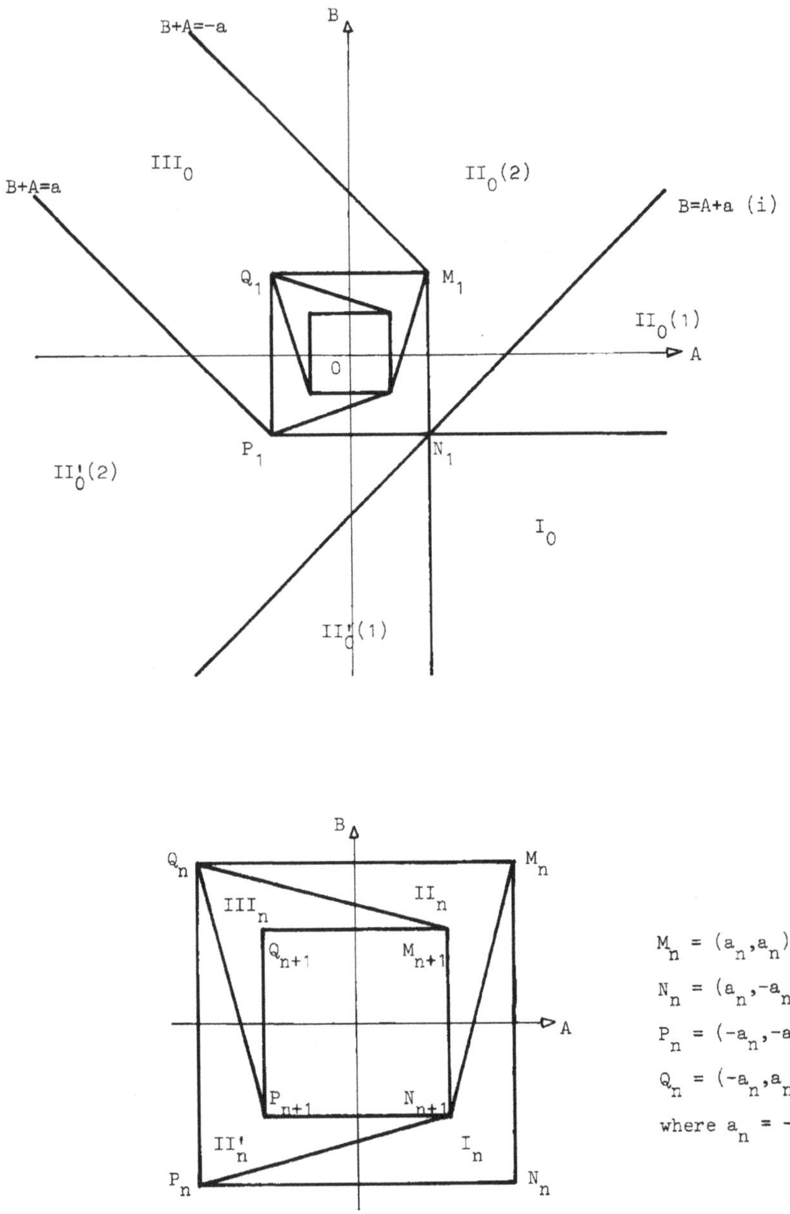


$M_n = (a_n, a_n)$

$N_n = (a_n, -a_n)$

$P_n = (-a_n, -a_n)$

$Q_n = (-a_n, a_n)$

where $a_n = -a/2n$

FIG. 15. Boundary value portrait of $P(a, \epsilon)$, $a < 0$ (the corresponding solutions are described in theorem 5).

viii) <u>if</u> $(A, B) \in \mathrm{III}_n$, <u>there exists</u> $4n+3$ <u>solutions</u>, <u>which are</u> $x_1^k$ <u>for</u> $k = 0, \dots, n-1$ <u>and</u> $x_2^k$, $x_3^k$ <u>and</u> $x_4^k$ <u>for</u> $k = 0, \dots, n$.

ix) <u>if</u> $(A, B) \in (M_n\ Q_n\ P_n) \cup (M_n\ N_{n+1}\ P_n)$ (<u>resp</u>. $(A, B) \in (Q_n\ M_{n+1}\ N_{n+1}) \cup (Q_n\ P_{n+1}\ N_{n+1})$ ), <u>there exists</u> $4n$ (<u>resp</u>. $4n+2$) <u>solutions which are</u> $x_1^k$, $x_2^k$, $x_3^k$ <u>and</u> $x_4^k$ <u>for</u> $k = 0, \dots, n-1$ (<u>resp</u>. $x_1^k$, $x_4^k$ <u>for</u> $k = 0, \dots, n-1$ <u>and</u> $x_2^k$, $x_3^k$ <u>for</u> $k = 0, \dots, n$ ).

9) <u>Remarks</u>.

i) In our discussion (e.g. uniqueness of the solutions in theorem 1 and section 8), two solutions of $P(a, \varepsilon)$ were considered equal if they have the same geometric shadow ; several families of solutions with the same limiting behaviour may exist (this is the case in section 8 for the $x_i^k$ ). Thus, although we don't distinguish such solutions, their asymptotic behaviour is entirely described. In a classical treatment, such non uniqueness cases are complicated due to the fact that solutions are studied as families $x(\varepsilon, t)$ depending on $\varepsilon$.

ii) Now, what about uniform approximates of the solutions of problem $P(a, \varepsilon)$ ? Assume $B - a > A > 0$ and let $x^i(t) = (B-a) \tanh \frac{1}{2}\left[\frac{B-a}{\varepsilon} t + \operatorname{Log} \frac{B-a+A}{B-a-A}\right]$ (resp. $x^\circ(t) = at + B - a$ ) be the solution of

$$\begin{cases} \dot{x} = \frac{1}{\varepsilon}\left[\frac{(B-a)^2}{2} - \frac{x^2}{2}\right] \\ x(0) = A \end{cases} \qquad \left(\text{resp. } \begin{cases} \dot{x} = a \\ x(1) = B \end{cases}\right).$$

We claim that

$$x(t) \sim x^c(t) = x^i(t) + x^\circ(t) - (B-a) \quad \text{on} \quad [0, 1] :$$

Indeed, change $t$ in $T = t/\varepsilon$ in $X_2(a, \varepsilon)$ ; you get the vector field $\tilde{X}_2(a, \varepsilon)$ of components $(u - \frac{x^2}{2}, \varepsilon ax)$.

For any finite $T$, its integral curve with initial conditions $(A, u_o)$ is infinitely close to that of $\tilde{X}_2(0, \varepsilon)$ with initial conditions $(A, {}^\circ u_o)$. Recall that ${}^\circ u_o = \frac{(B-a)^2}{2}$ for the solution of $P(a, \varepsilon)$ (see section 4.i for the case $a > 0$ and 7.ii for the case $a < 0$ ) ; then $x(t) \sim x^i(t)$ for any $t$ with $t/\varepsilon$ finite. Moreover $x(t) \sim x^\circ(t)$ for any $t$ with $t/\varepsilon$ infinitely large. Then, applying the permanence principle and the fact that $x^i(t) \sim B - a$

(resp. $x°(t) \sim B-a$ ) for any $t$ with $t/\varepsilon$ infinitely large (resp. for any infinitesimal $t$ ), we get the uniform approximate $x(t) \sim x^c(t)$ on $[0,1]$. Other boundary layers characters may be handled similarly. In case of a free layer, e.g. $a>0$, $B-A>a$ and $|A+B| < a$ (see theorem 4v), we consider

$$x^i(t) = \frac{B-A-a}{2} \tanh \tfrac{1}{2} \left[\frac{B-a}{\varepsilon}(t-t_*) + \operatorname{Log} \frac{B-A-a+2x_*}{B-A-a-2x_*}\right],$$

where $t_* = \frac{1}{2} - \frac{A+B}{2a}$ and $x_* = {}°x(t_*)$, the solution of

$$\begin{cases} x'(t) = \frac{1}{\varepsilon}\left(\frac{(B-A-a)^2}{8} - \frac{x^2}{2}\right) \\ x(t_*) = x_* \end{cases}$$

As previously we infer that $x(t) \sim at + x^i(t)$ on $[0,1]$. Here there is some ambiguity because we didn't know $x_*$ (it is conjectured that $x_* = 0$ ) ; however we can locate the free layer, because any possible value of $x_*$ (i.e. $|x_*| < \frac{B-A-a}{2}$ see section 4.iii) leads to the same geometric shadow for the solution $x(t)$. This shows the versality of our technic in these problems. In the method of matched asymptotic expansions (see, e.g., [F] for a precise formulation of this matter) you have to compute an inner and an outer expansions (the functions $x^i$ and $x^°$ considered above are the first terms of these expansions) and make the crucial work to verify that they both hold in an overlap region where they can be matched. For our purpose, we don't a priori need such expansions and may a posteriori get the composite expansion $x^c(t)$ by matching $x°(t)$ and $x^i(t)$ without the use of a strong overlaping assumption.

10) Exercises.

i) Prove theorems 3, 4 and 5 by observations in the stretched phase plane $(x, y = \varepsilon x')$. What is the behaviour of the shadows of the integral curves ? Compare with the Lienard plane.

ii) Prove that the approximates of the solutions of $P(a,\varepsilon)$ given in theorems 3, 4 and 5 are valid at any order $\varepsilon^n$ ( $n$ finite) by means of the change of variable $w_n = \frac{v-a}{\varepsilon^n}$. Using the change of variable $w = (v-a)^{[\varepsilon]}$, where $(v-a)^{[\varepsilon]}$ is the odd function of $v-a$ equal to $(v-a)^\varepsilon$ when $v-a$ is positive, we can get this property. What is the behaviour of the shadows of the integral curves in the plane $(x, w)$ ? Note that this very interesting change of variable was successfully used by several authors to describe precisely (see [BCDD], [D]) the integral curves in two-parameters singular perturbation problems.

11) References.

[BCDD] BENOIT E., CALLOT J.L., DIENER F. et DIENER M. : Chasse au canard. Publication IRMA (1980). Strasbourg.

[Bu] BURGERS J.M. : A mathematical model illustrating the theory of turbulence. Adv. Appl. Mech. 1 (1948).

[Coc] COCHRAN J.A. : Problems in singular perturbation theory. Doctoral dissertation (1962). Stanford University.

[Col] (*)

[D] DIENER F. : Les canards de l'équation $y'' + (y' + a)^2 + y = 0$ . Publication IRMA (1980). Strasbourg.

[DPS] DORR F.W., PARTER S.V. and SHAMPINE L.F. : Application of the maximum principle to singular perturbation problems. SIAM Review 15 (1973), p. 43-88.

[F] FRAENKEL L.E. : On the method of matched asymptotic expansions. Proc. Camb. Phil. Soc., 65 (1969), p. 209-284.

[Ha] HARRIS W.A. Jr. : Applications of the method of differential inequalities in singular perturbations problems, in New Developments in Differential Equations (W. Eckhaus ed.). North-Holland Publishing Company (1976).

[Ho] HOWES F.A. : Boundary-interior layer interactions in nonlinear singular perturbation theory. Mem. Amer. Math. Soc. 15, n° 203 (1978).

[M] MURRAY J.D. : On Burgers'model equation of turbulence. J. Fluid. Mech. 59 (1973), p. 263-279.

(*) [Col] COLE J.D. : Perturbation methods in applied Mathematics, Blaisdell, Waltham, Mass., 1968.

# Lesson 13

## A SPORTSMAN STORY

Problem. Describe the asymptotic behaviour of solutions in the two point problem $P_\varepsilon : \varepsilon^2 x'' = F(x)$ , $0 \le t \le 1$ , $x(0)$ , $x(1)$ prescribed, as $\varepsilon$ tends to $0$ .

THEOREM. The complete description may be performed by phase plane observation, at least when $F$ has a finite number of zeros. There are multiple solutions in most cases, with all kinds of layers. The location of the transition layers only depends on the first non-zero derivatives of $F$ at its zeros.

Comments. 0) Precise statements about the theorem will appear in the text, for the matter is too complicated to be told in few words !

The problem has highly significant applications in physics (non linear spring motion, solitary waves, diffusion in chemical reactions) which make its solution really worthwhile.

It has also some technical pecularities, discovered by G.F. CARRIER and C.E. PEARSON, which make the formal asymptotic approximation method inapplicable, namely the existence of formal solutions which correspond to no actual ones !

Also notice that there is a gap between the non autonomous problem $\varepsilon^2 \ddot{x} = F(t , x)$ and the present one ; in an unpublished paper by P.C. FIFE, we find a solution with a transition layer in case $F(t , x)$ depends non trivially on $t$ ; this layer depends on $F$ in a completely different way as in the autonomous case (we shall explain why in lesson 14).

A last important remark has to do with polemic about phase plane arguments used as a classical tool in asymptotics. Let us tell it in a picturesque manner !

As we prepared this lesson after [L.S.], we got a big shock as we found the problem solved in a paper by R.E. O'MALLEY four years ago, using phase plane arguments of the type we are accustomed to ! This time, N.S.A. was beaten on a ground where it seemed to be the best tool (in our mind, indeed...). Even pictures were used by a classical specialist in asymptotics, what is rather uncommon, for pictures don't help when you compute a formal approximation or when you prove its closeness with an actual solution as $\varepsilon$ is little.

Thus our conviction that formal approximation was nearly the only valuable technic in the classical treatment of singular perturbations has to be revisited : phase plane arguments may also be used, all the more as the first technic is forbidden here !

Indeed, O'MALLEY's paper has a strong "infinitesimal" flavour, due to its intuitive style that seems easy to formalize in "$\varepsilon - \delta$"-words, or within N.S.A. Thus N.S.A. appears as nothing more than a correct foundation for intuitive arguments, a mere trifle for applied mathematician ..

Then we had an eager discussion with LUTZ and SARI, as it became clear that their results on the layer's location were not exactly the same as O'MALLEY's ; this certainly was a bad catchword for N.S.A., and in the future, they had to make their computations sure before claiming their answers, etc...

However, we had an illfeeling about intuitive arguments : we believed that they could not succeed in proving such delicate results like free layer's location or oscillating jumps description, which concern the unexplicit relation between $\varepsilon$ and the solution of $P_\varepsilon$. Of course, a good intuition of the problem may avoid big mistakes - and this is the case in O'MALLEY's paper - but is not sufficient to avoid any mistake, and indeed, we found that the location of layers was not accurate.

At this stage, we have to discuss in details the pitfall on some example, before going on with a non standard treatment of the problem.

1) Consider a $C^\infty$ function $f : \mathbb{R} \longrightarrow \mathbb{R}$ with three singular points $\alpha < \beta < \gamma$, a maximum between two minima ; suppose $f(\alpha) = f(\gamma) = 0$ and look for solutions of $2\varepsilon^2 \ddot{x} = f'(x)$ such that $x(0) = a < \alpha$ and $x(1) = b > \gamma$.

We have energy conservation $(\varepsilon \dot{x})^2 - f(x) = k$ along any integral curve and this suggests observation in the stretched phase plane $(x, y)$ where $y = \varepsilon \dot{x}$. We get system

$$\begin{cases} \dot{x} = \dfrac{y}{\varepsilon} \\ \dot{y} = \dfrac{F(x)}{2\varepsilon} \end{cases}$$

and the energy $y^2 - f(x)$ is minimal at $(\beta, 0)$, saddle-point like at $(\alpha, 0)$, $(\gamma, 0)$.

The separatrix $\Gamma$ between closed equipotentials round $(\beta, 0)$ and those of level under $f(\alpha) = 0$ has equation $y = \pm\sqrt{f(x)}$. We get the following picture :

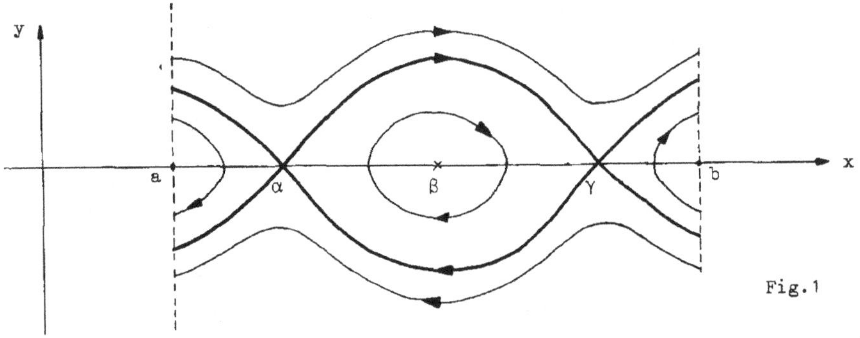


Fig.1

For $k < 0$, there is no solution from $a$ to $b$ ; all curves with $k > 0$ in the upper half plane join $a$ to $b$ in the time $\tau(k) = \int_a^b \frac{\varepsilon \, dx}{\sqrt{f(x) + k}}$ ; as $\tau(k)$ is decreasing from big values for little $k$ to little values for big $k$, there is exactly one $k_\varepsilon$ with $\tau(k_\varepsilon) = 1$, that is an unique solution $x_\varepsilon(t)$ for our problem. Moreover $k_\varepsilon = O(\varepsilon^2)$ ; thus the curve $(x_\varepsilon(t), y_\varepsilon(t))$ goes along the upper part of $\Gamma$ on a level $O(\varepsilon^2)$, with speed of order $\frac{1}{\varepsilon}$ outside little neighbourhoods of $\alpha$ and $\gamma$. In other words, we have two boundary layers and a transitive layer, all of thickeness $O(\varepsilon)$. To locate the inner layer, remark that the large contributions to $\tau(k_\varepsilon)$ occur near $\alpha$ and $\gamma$, the points away from these having asymptotically negligible contributions.

Further, since $f(x) \approx \frac{f''(\alpha)}{2}(x-\alpha)^2$ near $\alpha$ and $f(x) \approx \frac{f''(\gamma)}{2}(x-\gamma)^2$ near $\gamma$ (provided $f''(\alpha)$ and $f''(\gamma)$ are nonzero), the proportionnal contributions are asymptotically $\frac{1}{\sqrt{f''(\alpha)}}$ and $\frac{1}{\sqrt{f''(\gamma)}}$. Hence the layer is located at time

$t_o = \dfrac{\sqrt{f''(\gamma)}}{\sqrt{f''(\alpha)}+\sqrt{f''(\gamma)}}$ . More generally, if $\alpha$ and $\gamma$ are of order $2n$ , the same argument shows that $t_o = \dfrac{\sqrt{f^{(2n)}(\gamma)}}{\sqrt{f^{(2n)}(\alpha)}+\sqrt{f^{(2n)}(\gamma)}}$ , and we have the following picture, which describes the non uniform limit of $x_\varepsilon$ in $[0,1]$ :

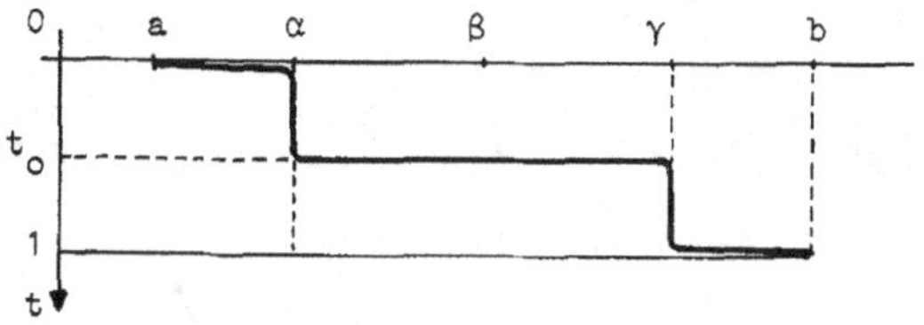


Fig.2

This is essentially O'MALLEY's result on this example: We shall prove that

$$t_o = \frac{[f^{(2n)}(\gamma)]^{\frac{1}{2n}}}{[f^{(2n)}(\alpha)]^{\frac{1}{2n}} + [f^{(2n)}(\gamma)]^{\frac{1}{2n}}} ,$$

which is not the same whenever $n>1$ .

Thus, either we have proved that $0=1$ , which after lesson II.1 has unpleasant effects on our faith in mathematics, either something has to be revisited...

Indeed, the intuitive proof above performs the canonical and ritual crime against infinitesimal calculus : we wrote $\dfrac{1}{\sqrt{\frac{f(x)}{\varepsilon^2}+\frac{k}{\varepsilon^2}}} \sim \dfrac{1}{\sqrt{\frac{f''(\alpha)(x-\alpha)^2}{\varepsilon^2}+\frac{k}{\varepsilon^2}}}$ , with $\frac{f(x)}{\varepsilon}$ little near $\alpha$ . In other words, we used the rule "from $v \sim 0$ , deduce $\frac{1}{u+v} \sim \frac{1}{u}$ ", which is false if $u \sim 0$ (write $\frac{1}{u+v} - \frac{1}{u} = \frac{v}{u(u+v)} \sim \frac{0}{0}$ ).

Therefore developping $f$ near $\alpha$ cannot be used as an argument, for we make approximations in the $(x,y)$-plane, which are only valuable for fixed $\varepsilon$ ; such approximations have no asymptotical meaning as $\varepsilon$ tends to $0$ !

2) We return now to the general problem and begin the non standard treatment with the usual preliminary transfer ; now $F$, $a$, $b$ are supposed to be standard. We assume that, in this discussion, $F$ has a finite set of zeros ; all are standard, of course.

Recall that any information about $P_\varepsilon$ , which is true for every $\varepsilon \sim 0$ translates into a classical " ε-little enough" statement.

Therefore, we fix $\varepsilon \sim 0$ $(\varepsilon > 0)$ and observe the integral curves in the $(x, y)$-plane as above. Call $f$ a fixed standard primitive of $2F$ and you have again conservation of energy $y^2 - f(x) = k$ . Call $Z$ the vector field of components $(\frac{y}{\varepsilon}, \frac{F(x)}{\varepsilon})$ . Outside the halos of its singular points, $\|Z\|$ is infinitely large ; this is an open door for quick jumps !

Assume that the zeros of $F$ are minima $\alpha_1, \dots, \alpha_p$ of $f$ alternating with maxima $\beta_1, \dots, \beta_q$, with $q = p \pm 1$ , as pictured below.

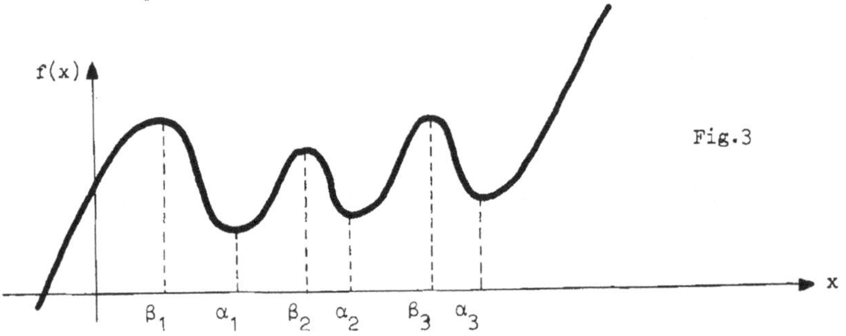


Fig.3

Later on, we shall assume all the $\alpha_i$ 's of standard orders $2n_i$ , that is the first non zero derivative at $\alpha_i$ is $f^{(2n_i)}(\alpha_i) > 0$ .

Call $\Gamma_i$ the curve of equation $y^2 = f(x) - f(\alpha_i)$ and $\Gamma = \bigcup_i \Gamma_i$ . These $\Gamma_i$ are separatrix curves in the phase plane portrait ; the whole qualitative behaviour of integral curves only depends on the signs of $f(\alpha_i) - f(\alpha_j)$ , which give the relative positions of the $\Gamma_i$ 's .

We picture all cases with $p + q \leq 4$ ; nothing essential would appear for larger values of $p + q$ :

p+q=1

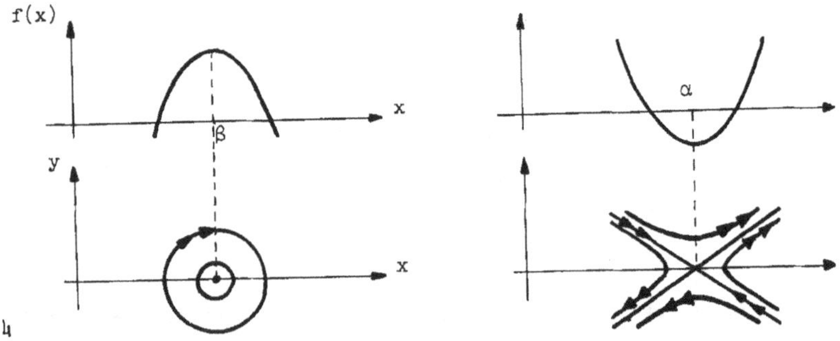


Fig. 4

p+q = 2

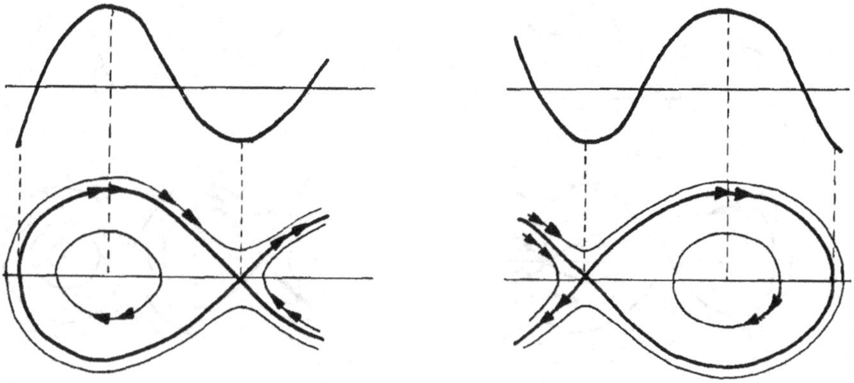

p+q = 3

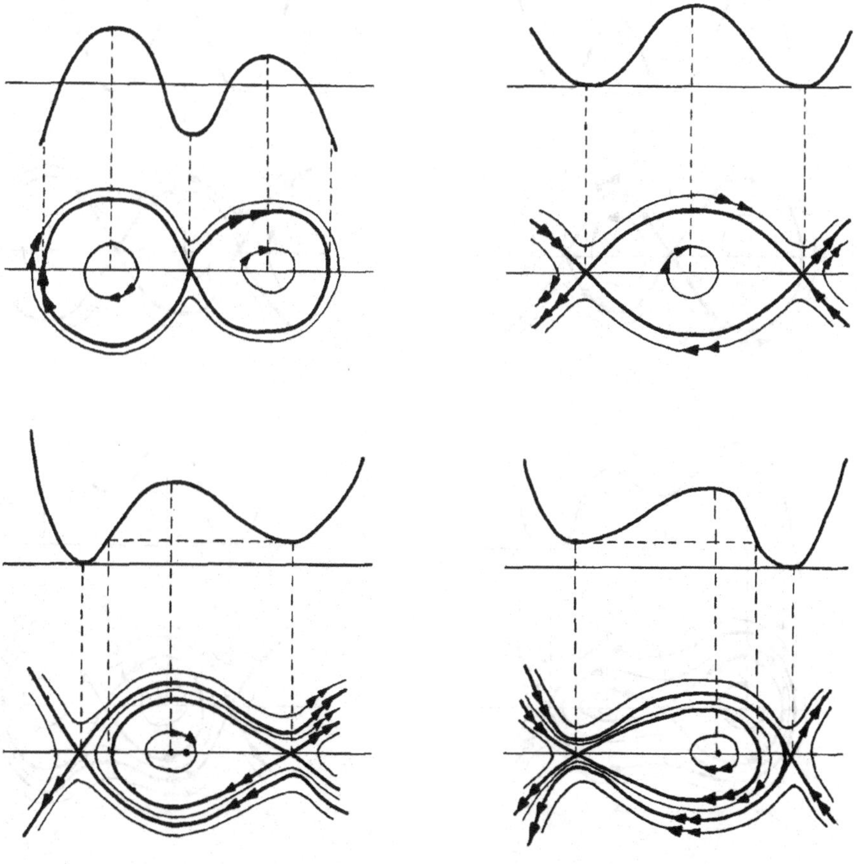

p+q = 4

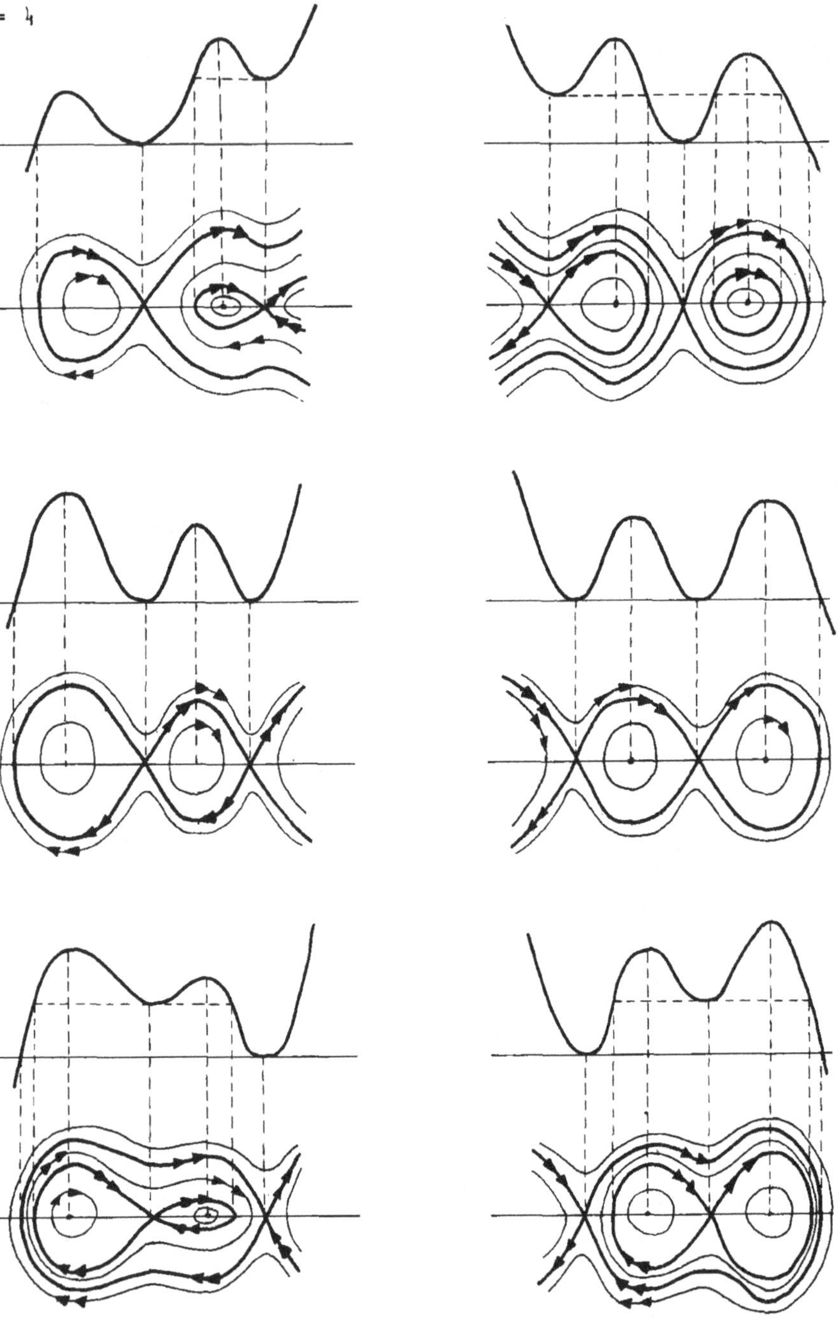

Note that if $f''(\alpha_i) \neq 0$ , the separatrix $\Gamma_i$ makes a regular cross at $(\alpha_i , 0)$ while if $f''(\alpha_i) = 0$ , we have horizontal tangents at $(\alpha_i , 0)$ . Also note that symmetric pictures give different solutions for $P_\varepsilon$ , if $a = x(0)$ and $b = x(1)$ are fixed.

3) Now consider an eventual solution $(x(t) , y(t))$ of problem $P_\varepsilon$ . This curve goes in time 1 from a point $u_o$ of the vertical $V_a$ to a point of the vertical $V_b$ . The only cases where it meets $\Gamma$ is whenever $a = b =$ some singular point : in these cases we have a trivial solution $x(t) = a$ , $y(t) = 0$ . We exclude these solutions in our discussion ; then equipotentials meet $V_a$ either transversally, either with a contact of order 1 on the x-axis. Therefore the $t \in [0 , 1]$ such that $(x(t) , y(t)) = u_o$ are isolated ; we call their number the <u>degree of the solution</u>. It is an integer that may be finite, i.e. standard, or infinitely large. In the first case $x(t)$ has finite variation if $u_o$ is finite (classically, the corresponding family of solutions has bounded variations as $\varepsilon$ tends to $0$ ), but finite variation may also occur with infinitely large degree (for instance, if $a = b = \beta$ , consider solutions in the halo of $(\beta , 0)$ ). Notice that solutions of infinitely large degree correspond to families $x_\varepsilon(t)$ , which have a limit on no fixed subinterval of $[0 , 1]$ , while finite degree insures limits with a finite number of non uniformities, as we shall see. Also solutions with $\|u_o\|$ infinitely large may occur in some cases (for instance if $p = 1$ , $q = 0$ or whenever $\Gamma$ is bounded) ; they correspond to families $x_\varepsilon(t)$ which are not bounded as $\varepsilon$ tends to $0$ .

We call of <u>finite type</u> all solutions of finite degree with $u_o$ finite. They move in the halo of $\Gamma$ , for outside the speed would be infinitely large along a curve of finite length ; moreover, their shadow contains at least one saddle point, for the halo's of these points are the only places where non infinitesimal time can be lost with finite variation.

This yields the following description, which seems intuitive, but is a precise (and proved !) statement :

<u>every solution of finite type jumps along</u> $\Gamma_i$ <u>into the halo of some</u> $(\alpha_i , 0)$ ,

then jumps from saddle point to saddle point along $\Gamma_i$ until a last jump leads to the arrival line $V_b$ . If $a$ or $b$ are at an $\alpha$ , the corresponding jump fails. The speed along the jumps is at least of order $\frac{1}{\varepsilon}$ ; therefore the boundary layers and transition layers associated with the jumps are of thickeness order $\varepsilon$ ; they are separated by a finite number of slow motions near saddle points.

Thus, to get all such solutions, we must recognize all possible shadows on the qualitative phase portrait: More precisely, let us call solution-scheme the following data :

i) a standard sequence $u_1, \dots, u_n$ of saddle points,

ii) a point $u_0 \in \Gamma \cap V_a$ and a point $u_{n+1} \in \Gamma \cap V_b$ ,

iii) for every $i$ , an oriented simple arc of $\Gamma$ from $u_i$ to $u_{i+1}$ , which contains no other singular point. Call $\gamma$ the corresponding oriented geometric curve from $u_0$ to $u_{n+1}$ ,

iv) a standard interval $I_a$ of $V_a$ with endpoint $u_0$ and a standard interval $I_b$ of $V_b$ with endpoint $u_{n+1}$ such that there exists an integral curve of $Z$ which has $\gamma$ as oriented geometric shadow and joins a point in $I_a$ to a point in $I_b$ . Call it a "test curve"

This is quite a monthful, as often in geometric descriptions. But some pictures will make things clear (the dotted lines represent test curves) :

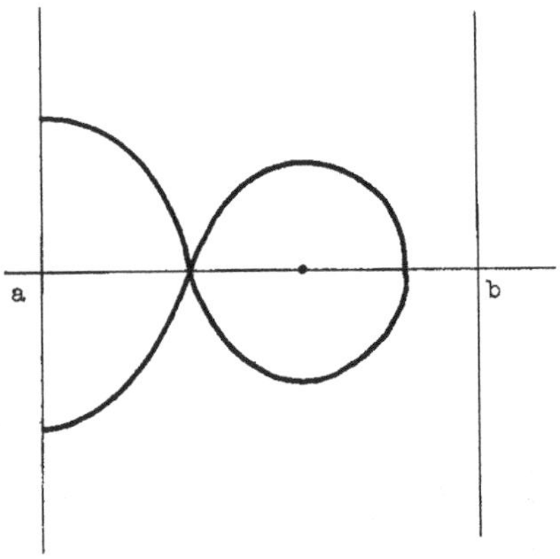


no solution at all

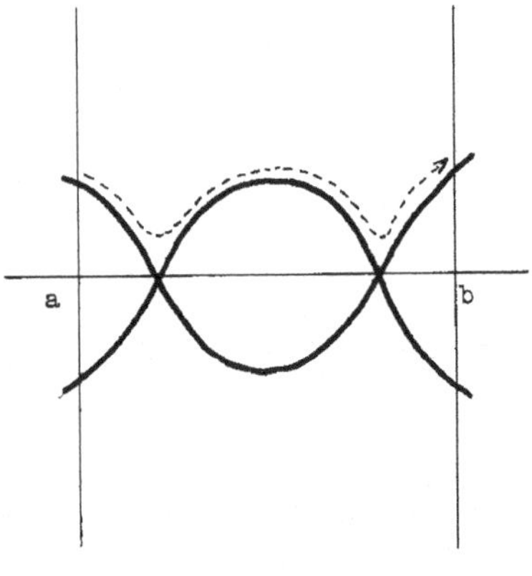


1 scheme ( of degree 1 ) .

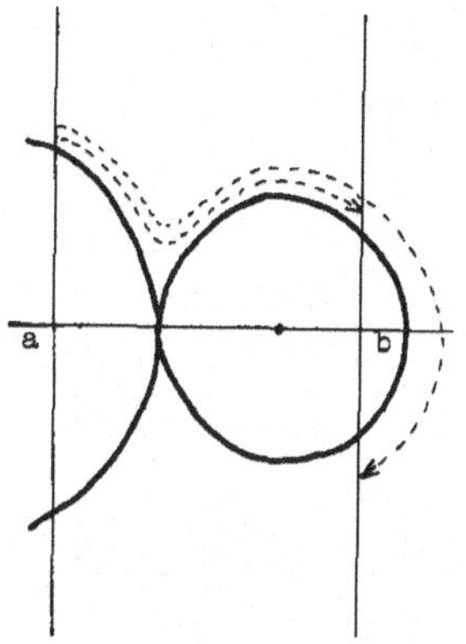


2 shemes (of degree 1)

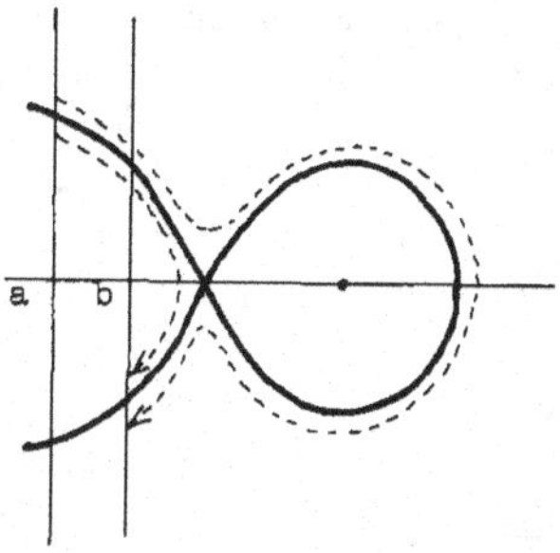


2 schemes (of degree 1)

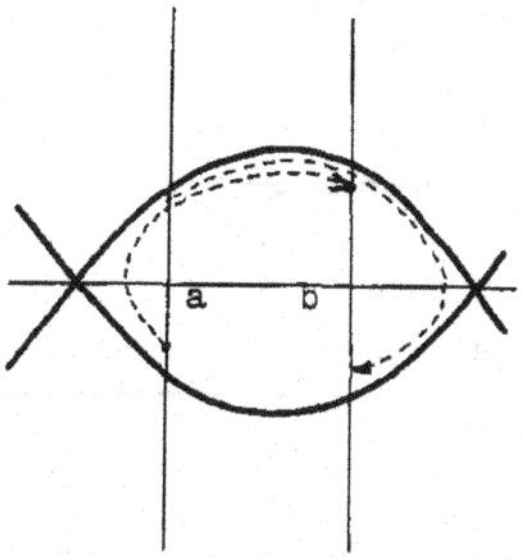


2 schemes of degree 1
8 " " 2
a lot of schemes of any degree

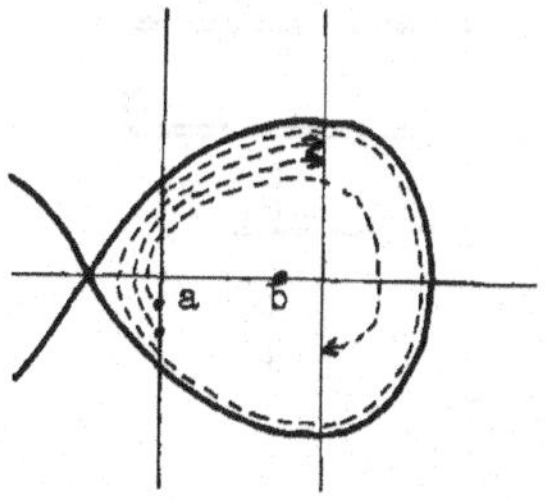


2 schemes of degree 1
8 " " 2
a lot of schemes of any degree

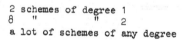

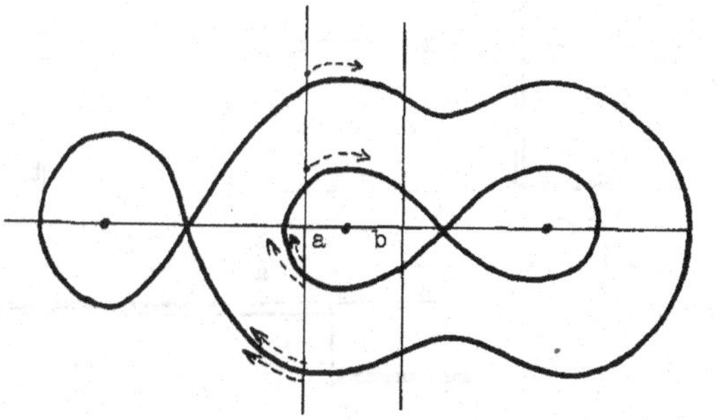


find the 7 schemes of degree 1 and the 3 kinds of schemes of each degree .

To every solution of finite type is associated a solution scheme. Conversely, <u>for every solution scheme</u>, <u>there is one and only one solution</u> (<u>of finite type</u>) <u>which has</u> $\gamma$ <u>as geometric oriented shadow and joins</u> $I_a$ <u>to</u> $I_b$ .

<u>Proof</u> (outline). It is clear that any integral curve starting at $(a, y) \in I_a$ with $(a, y) \sim u_o$ meets $I_b$ and has $\gamma$ as oriented geometric shadow, for it follows the test curve. The time spent depends continuously and monotoneously on the energy level, that is on $y$ . If $y$ is far enough, this time is $\sim 0$ (use the permanence principle) ; if $(a, y)$ tends to $u_o$ , this time tends to infinity, for the curve approaches at least one singular point. Hence there is exactly

one solution with the prescribed scheme .

et us draw some pictures of solutions in the $( t , x )$ - plane :

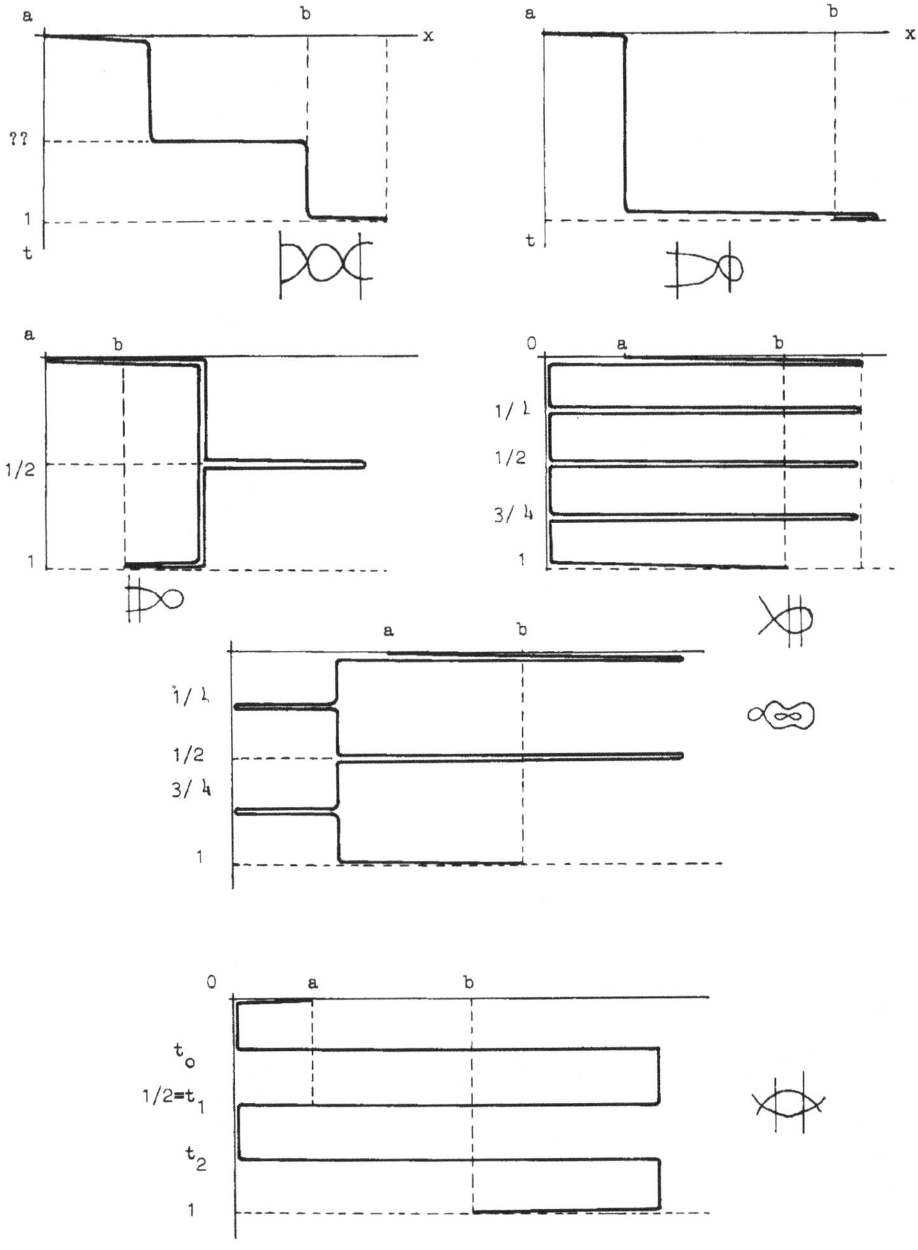


$t_o = t_2 - t_1$ et $t_1 - t_o = 1 - t_2$ thus $t_1 = 1/2$ . But $t_o$ = ???

As we see, in some cases the transition layers are located by symmetry, or periodicity arguments. But whenever the solution passes by different saddle points, the relative time spent near them may be also different and we have to compare them. Before doing this, let us briefly consider solutions of infinitely large degree with $\|u_o\|$ finite.

4) Such solutions have as geometric support a closed equipotential which meets both $V_a$ and $V_b$. On such a curve, the time spent to turn once is some number $\tau$ of order $\varepsilon$, except for some cases, when turning in the halo of a center (this needs $a = b$).
Now there is an integer such that $p\tau \leq 1 < (p+1)\tau$ and it is easy to check with arguments as above that there is exactly one solution of degree $p$ turning in the halo of $\gamma$ with starting and arrival places specified as regards the sign of $y(0)$ (if not, we get 4 solutions).
Thus we have solutions which oscillate a number of times of order $\frac{1}{\varepsilon}$. In spite of the absence of a "limit", such solutions are important in applications...

5) <u>The location of layers for finite type solutions</u>. Consider a solution scheme with saddle points $(\alpha_1, 0), \ldots, (\alpha_n, 0)$ and the corresponding solution $(x(t), y(t))$ with energy $k = y^2 - f(x)$. Assume that each $\alpha_i$ is a zero of finite order for $f'$, that is $f^{(2p_i)}(\alpha_i) \neq 0$ and $f^{(m)}(\alpha_i) = 0$ for $1 \leq m < 2p_i$ (recall that $f(\alpha_i)$ is minimal).
The transition layers are located in the halos of the standard points $t_i$, with $t_o = 0 \leq t_1 \leq \ldots \leq t_{n-1} \leq t_n = 1$ such that on every standard subinterval of $]t_{i-1}, t_i[$, one has $x(t) \sim \alpha_i$ (see lesson 10) ; in other words, $t_i$ is the common shadow of the times where $(x(t), y(t))$ is on the jump from $\alpha_i$ to $\alpha_{i+1}$, but not in the halo of $\alpha_i$ and $\alpha_{i+1}$.
Notice that we consider $t_i = t_{i-1}$ as concerning <u>two</u> different jumps ; the time spent near $\alpha_i$ is infinitesimal, but for engineers, a jump with speed of order $\frac{1}{\varepsilon}$ along it and a double jump with a quick slackening in the middle under order $\frac{1}{\varepsilon}$ have not the same meaning.

THEOREM. *Call* $p = \sup p_i$ *and* $I = \{1 \le i \le n\ ,\ p_i = p\}$ . *Then*

$$L = \sum_{i \in I} \frac{1}{[f^{(2p)}(\alpha_i)]^{\frac{1}{2p}}}$$

*is defined and we have, provided* $a \ne \alpha_1$ *and* $b \ne \alpha_n$ :

i) *if* $i \notin I$ , *then* $t_i = t_{i-1}$ ,

ii) *if* $i \in I$ , $t_i - t_{i-1} = \frac{1}{L}[f^{(2p)}(\alpha_i)]^{1/2p}$ .

*Moreover, if* $a = \alpha_1$ *and* $p_1 = p$ , *replace* $[f^{2p}(\alpha_1)]^{1/2p}$ *by its half. Same rule if* $b = \alpha_n$ *and* $p_n = p$ .

Thus, the point $(x(t)\ ,y(t))$ spends the best of the time at the most conforta-ble minima. This epicurean behaviour is rather natural, indeed... What a fascina-ting equation, isn't it ?

Note that the exponent $\frac{1}{2p}$ is mostly different from $\frac{1}{2}$ .

*Proof.* We have to estimate up to an infinitesimal the time spent by $(x(t)\ ,y(t))$ in some standard square containing the halo of a saddle point $(\alpha\ ,0)$ and no other one.

There are two cases, according to the sign of $k + f(\alpha)$ :

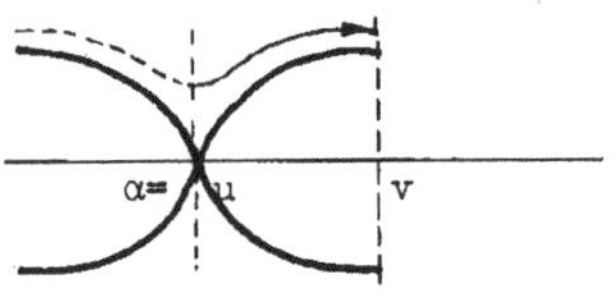


$k + f(\alpha) > 0$

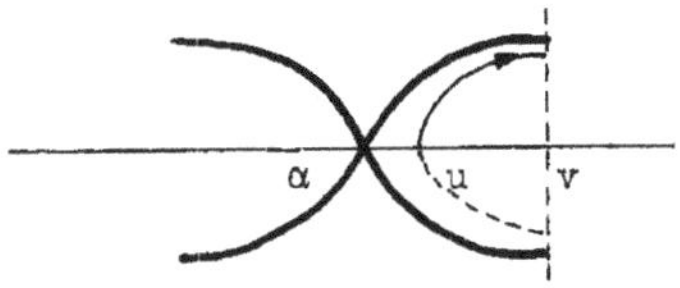


$k + f(\alpha) < 0$

Cutting in two halves and using symmetries, we reduce the problem to the time es-timate between $x = u$ and $x = v$ for $v$ standard and $v > \alpha$ , as pictured above. We know a priori that this time $\tau$ is at most 1, for $(x(t)\ ,y(t))$ is a solution of $P_\varepsilon$ , and that the choice of $v$ affects $\tau$ only up to an infinitesimal. Let $2q - 1$ be the order of $\alpha$ as a root of $f'$ . Then from $f(u) \sim f(\alpha)$ , we deduce $u \sim \alpha$ , by Taylor's formula.

Now, from $\dot{x} = \frac{y}{\varepsilon}$ and $y = \sqrt{f(x) + k}$ , we get $\tau = \int_u^v \frac{\varepsilon\, dx}{\sqrt{f(x)+k}}$ , which in both cases is $\lim_{w \to u} \tau(w, f)$, where $u < w < v$ and $\tau(w, f) = \int_w^v \frac{\varepsilon \, dx}{\sqrt{f(x) + k}}$. Further suppose $f(\alpha) = 0$ (replace $f$ by $f - f(\alpha)$) and consider, for some $r \leq q$, the standard mapping $\varphi : [\alpha, v] \longrightarrow [\alpha, \bar{v}]$ such that $\varphi(x) = \alpha + [f(x)]^{1/2r}$, with $\bar{v} = \varphi(v)$. It is $C^\infty$ on $]\alpha, v]$ if $f$ is $C^\infty$ and

$$\varphi'(x) = \frac{1}{2r} [f(x)]^{(1/2r)-1} f'(x) > 0 .$$

LEMMA 1. i) <u>if</u> $r < q$, $x \sim \alpha$ <u>implies</u> $\varphi'(x) \sim 0$;

ii) <u>if</u> $r = q$, $x \sim \alpha$ <u>implies</u> $\varphi'(x) \sim \left[\frac{f^{(2q)}(\alpha)}{(2q)!}\right]^{\frac{1}{2q}}$.

Indeed, from Taylor's formula applied to $f(x)$ and $f'(x)$ we get, for $x \sim \alpha$,

$$\varphi'(x) = \frac{1}{2r} \left[\frac{(x-\alpha)^{2q}}{2q!} (f^{(2q)}(\alpha) + \eta)\right]^{\frac{1}{2r} - 1} \times \frac{(x-\alpha)^{2q-1}}{(2q-1)!} (f^{(2q)}(\alpha) + \nu) ,$$

with $\eta \sim \nu \sim 0$. Hence $\varphi'(x) = K(x-\alpha)^{\frac{2q}{2r} - 1} ([f^{(2q)}(\alpha)]^{\frac{1}{2r}} + \mu)$, with $K = \frac{1}{2r[(2q)!]^{(1/2r)-1} \times (2q-1)!}$ and $\mu \sim 0$. This proves lemma 1.

LEMMA 2. <u>Assume</u> $w \sim \alpha$. <u>Then</u>.

i) <u>if</u> $r < q$, $\bar{\tau} = \int_{\varphi(w)}^{\varphi(v)} \frac{\varepsilon \, dz}{\sqrt{(z-\alpha)^{2r} + k}} \sim 0$;

ii) <u>if</u> $r = q$, $\bar{\tau} = \int_{\varphi(w)}^{\varphi(v)} \frac{\varepsilon \, dz}{\sqrt{(z-\alpha)^{2q} + k}} \sim \left[\frac{f^{(2q)}(\alpha)}{(2q)!}\right]^{\frac{1}{2q}} \tau(w, f)$.

To prove this, we use the fact that $k$ is a number for which $\tau(w, f) \leq 1$. Indeed, $\bar{\tau} = \int_w^v \frac{\varepsilon \, \varphi'(x) \, dx}{\sqrt{f(x) + k}} \sim \int_w^s \frac{\varepsilon \, \varphi'(x) \, dx}{\sqrt{f(x) + k}}$, for every standard $s \leq v$, for on $[s, v]$, the integrant is infinitesimal. By permanence, this property is true on some $[s_0, v]$ with $w < s_0 < v$ and $s_0 \sim w$. For the same reason, there is an $s'_0 \sim w$ such that on $[s'_0, v]$, $\tau(w, f) \sim \int_w^s \frac{\varepsilon \, dx}{\sqrt{f(x) + k}}$. We fix $s \sim w$, $s \geq s_0$, $s \geq s'_0$. Then $0 \leq \int_w^s \frac{\varepsilon \, \varphi'(x) \, dx}{\sqrt{f(x) + k}} \leq \sup_{[w,s]} \varphi'(x) \int_w^s \frac{\varepsilon \, dx}{\sqrt{f(x) + k}}$; thus, if $r < q$, the last number is infinitesimal (use lemma 1 and $\tau(w, f) \leq 1$), which proves (i). If $r = q$, we have $\bar{\tau} \sim \varphi'(\xi) \int_w^s \frac{\varepsilon \, dx}{\sqrt{f(x) + k}}$ for some $\xi \in [w, s]$, hence

$\bar{\tau} \sim \varphi'(\xi)\ \tau(w\,,\,f)$ , which proves (ii) by lemma 1.

Passing to the limit as $w$ tends to $u$ , we get from lemma 2 the compared time estimates if $f$ is replaced by $(z-\alpha)^{2r}$ <u>for the same</u> $k$ . To prove the theorem, first notice that on a whole passage near $\alpha$ , the time estimates have to be taken twice, but if $a=\alpha_1$ (or $b=\alpha_n$ ), only one contribution occurs.

With lemma 2, we may compare the times $\tau_i$ and $\tau_j$ near $\alpha_i$ and $\alpha_j$ : if the orders are the same, we have

$$\left[f^{(2p_i)}(\alpha_i)\right]^{\frac{1}{2p_i}} \tau_i \sim \left[f^{(2p_j)}(\alpha_j)\right]^{\frac{1}{2p_j}} \tau_j \ ,$$

and if $p_i > p_j$ , then $\tau_j \sim 0$ .

Hence only the maximal ordered $\alpha_i$ have a non infinitesimal contribution $\tau_i$ :

As $t_i - t_{i-1} = {}^\circ\tau_i$ , this ends the proof.

Furthermore, we can nearly compute the energy $k$ along any solution of finite type. A first insight shows that it is certainly of order $\varepsilon^2$ . But this is a very rough estimate ; consider for instance some $\alpha$ of the highest order $2p$ in the "horizontal" case $k>0$ . The corresponding time

$$\tau \sim 2\left[\frac{f^{(2p)}(\alpha)}{(2p)!}\right]^{\frac{1}{2p}} \int_0^1 \frac{\varepsilon\, dz}{\sqrt{z^{2p}+k}} = 2h\ I(k) \ .$$

We know ${}^\circ\tau$ from the theorem and deduce $k$ from $I(k) \sim \frac{\tau}{2h}$ . Let us compute it for $p=1$ , the nondegenerate case.

We get $2I(k) = -2\varepsilon \log \dfrac{\sqrt{1+k}-1}{\sqrt{k}} \sim -\frac{\varepsilon}{2} \log k \sim \frac{\tau}{h}$ . Thus $k$ is of order type $\exp -\frac{K}{\varepsilon}$ with $K$ non infinitesimal. ( $K$ is any standard number with $K < {}^\circ(\frac{\tau}{h})$ .)

<u>Remark</u>. It seems rather cumbersome to give a classical proof of the theorem (but there is one, as we know from lesson II.7) ; you would have to replace our's in the proof of Lemma 2 by some suitable function of $\varepsilon$ and estimate integrals with moving bounds. A lot of work, of course...

6) The restriction we made about the number of roots of $F$ is not essential. You can use our technic in more general cases, e.g. $F(x) = \sin \pi x$ . It is easy to check that a finite type solution only passes near a standard number of roots and the study is the same as above.

Another generalization is to consider equation $\varepsilon(p(x)\dot{x})' = F(x)$ with $p(x)>0$ . But it reduces to the case $p(x)=1$ if we put $X=q(x)$ with $q'=p$ , for $q$ is inversible.

7) Now, what about the title of this lesson ? We didn't see any sportsmen here... Maybe ! But did you ever observe a cat running after a mouse along some straight line ? At time 0, the cat is at $a$ and the mouse (an uniformly running one !) will be at $b$ at time 1.

There are different hiding places $\alpha_i$ for the cat, where it may loose time: trees alternate with water holes, but the cat doesn't like water and has to jump from tree to tree in order to catch the mouse at time 1. The strong acceleration outside the trees is given by a force $\frac{F(x)}{\varepsilon}$ with $\varepsilon$ little, and we have problem $(P_\varepsilon)$ as a model for the cat's strategy. Note its accuracy with the natural tendancy to use the most comfortable "flat trees" as relaxation places... Also note in some cases the cunning behaviour of the cat, which goes first in the opposite direction or makes "failed jumps" from one tree to the same.

Of course, you have not to believe this sportsman story !

8) Exercise. Describe the asymptotic behaviour of solutions in problem

$$\varepsilon^2 \frac{\partial^2\varphi}{\partial t^2} = F(\varphi) \ , \quad \varphi(0, x) \ \text{and} \ \varphi(1, x) \ \text{prescribed},$$

where $\varphi(t, x)$ has to be defined on a square $[0,1]\times[0,1]$ .

9) Classical references. G.F. CARRIER and C.E. PEARSON. "Ordinary differential equations", Blaisdell, Walthram 1968.

R.E. O'MALLEY Jr. "Phase-plane solutions to some singular perturbation problems". J. Math. Anal. Appl. 54 (1976), pp 449-466.

Lesson 14

# FORCED LAYERS IN A NONAUTONOMOUS PROBLEM

Problem. Describe the asymptotic behaviour of the solutions in the two-point problem

$$(P_\varepsilon)\quad \begin{cases} \varepsilon^2(p(t,x,\varepsilon)x')' = F(t,x,\varepsilon) \\ x(0)=a\ ,\quad x(1)=b \quad \text{prescribed} \\ 0 < p(t,x) \le 1\ . \end{cases}$$

THEOREM. Under suitable conditions, there is a family $x_\varepsilon(t)$ of solutions with boundary layers and a transition layer at a prescribed time $t_0 \in ]0,1[$ . All these layers are of thickeness order $\varepsilon$ .

Comments. 0) This is a famous problem, which curiously was studied in the litterature before the autonomous case of lesson 13. Indeed, in a well-known paper (still unpublished, to our knowledge) on "Two point boundary value problems admitting interior transition layers", P.C. FIFE proved that under suitable conditions on P and F , there is a solution with inner transition layer at time $t_0$ jumping from one solution of the reduced equation $F(t,x,0)$ to another. (See also F.A. HOWES "Singular perturbations and differential inequalities", A.M.S. Memoirs 168 (1976)). These technical conditions are required in order to make a development procedure go (first, one describes a formal solution with the expected behaviour and then proves its closeness to an actual solution) ; the proof is surprisingly difficult, involving fine tools of analysis. Moreover, this technic seems unable to give some information about all asymptotic solutions of $(P_\varepsilon)$ .
In the autonomous problem of lesson 13, Fife's conditions are not all fulfilled, because they essentially need non trivial time dependance at least near $t_0$ , in order to force the inner layer to occur at time $t_0$ and not at some a priori unknown place as in the autonomous case. Thus it is not surprising that, as long as classical technics are used and no general solution of $(P_\varepsilon)$ is expected, the nonautonomous case with a priori forced layer location seems easier than the

autonomous case.

However, solutions with unexpected layers may occur, strongly depending on each other as in the autonomous problem, even if Fife's conditions are satisfied. Thus it seems not entirely true that, as suggested in his paper, the non trivial time dependance localizes the various layer phenomena, which should have very little effect on each other ; only the expected layers are localized by the conditions.

The aim of this lesson is to give a geometric meaning to the "mysterious" conditions of P. Fife, to outline an alternative proof of his theorem (so far as no asymptotic approximate is required), and a general treatment of the problem, involving the case of "unexpected layers".

1) Let us recall Fife's conditions, up to a slight change of notations. First write $\varepsilon^2(p(t,x,\varepsilon)\dot{x})^{\cdot}\, p(t,x,\varepsilon) = F(t,x,\varepsilon)\, p(t,x,\varepsilon) = H(t,x,\varepsilon)$ (this technical device has some tradition ; it yields a self adjoint operator which is easier to handle.).

<u>Conditions</u> :

($F_1$) $H(t,x,0)$ has at least two distinct solutions $g_1(t)$ and $g_2(t)$ on $[0,1]$, i.e. $H(t,g_i(t),0)=0$ . Assume $g_1(t)<g_2(t)$ .

($F_2$) Smoothness conditions up to order 2 for $p, H$ and boundedness of the second derivatives of $H$ for $\varepsilon$ little. Moreover $\frac{\partial H}{\partial x}(t, g_i(x), 0) \geq K > 0$ and $1 \geq p \geq K$ , where $K$ is a constant.

($F_3$) For each $k \neq g_1(0)$ in the closed interval between $g_1(0)$ and $a$ , one has $\int_{g_1(0)}^{k} H(0,x,0)\,dx > 0$ .

($F_4$) For each $k \neq g_2(1)$ in the closed interval between $g_2(1)$ and $b$ , one has $\int_{g_2(1)}^{k} H(1,x,0)\,dx > 0$ .

($F_5$) $\int_{g_1(t_o)}^{k} H(t_o,x,0)\,dx$ is $\begin{cases} > 0 & \text{for } k \in \,]g_1(t_o), g_2(t_o)[ \\ = 0 & \text{for } k = g_2(t_o) . \end{cases}$

($F_6$) $\int_{g_1(t_o)}^{g_2(t_o)} \frac{\partial H}{\partial t}(t_o,x,0)\,dx \neq 0$ .

<u>Conclusion</u>. For $\varepsilon$ little enough, there exists a family $x(t,\varepsilon)$ of solutions of

$(P_\varepsilon)$ such that for any $\omega > 0$ ,

$$\lim_{\varepsilon \to 0} x(t, \varepsilon) = \begin{cases} g_1(t) & \text{on } [\omega, t_o - \omega] \\ g_2(t) & \text{on } [t_o + \omega, 1 - \omega], \end{cases}$$

that is two boundary layers and a transition layer at time $t_o$ .

Now forget provisionally these conditions and try to solve the problem with your naive non-standard mind.

2) First transfer the problem as usual, i.e. take $a, b, F, p$ standard and consider equation $\varepsilon^2(px')' = F$ with $\varepsilon > 0$ , fixed. Assume $p > 0$ and introduce the auxiliary variable $y = \varepsilon p(t, x, \varepsilon)x'$ as in lesson 13. In the $(t, x, y)$-space you get the equivalent system

$$\begin{cases} x' = \dfrac{y}{\varepsilon p(t, x, \varepsilon)} \\ y' = \dfrac{F(t, x, \varepsilon)}{\varepsilon} \\ t' = 1 , \end{cases}$$

which suggest the "failed prime integral trick" $yy' = x'pF$ ; thus introduce $H(t, x, \varepsilon) = F(t, x, \varepsilon)\, p(t, x, \varepsilon)$ and you get

$$\begin{cases} x' = \dfrac{y}{\varepsilon p} \\ y' = \dfrac{H}{\varepsilon p} . \end{cases}$$

For any integral curve $\gamma(t) = (t, x(t), y(t))$ , we have

$$y^2(s) - y^2(t) = 2\int_t^s H(\tau, x(\tau), \varepsilon)\, x'(\tau)\, d\tau ,$$

which unfortunately is not as nice as in the autonomous problem. Let $f(t, x, \varepsilon)$ be some standard $C^1$ function such that $\frac{\partial f}{\partial x}(t, x, \varepsilon) = 2H(t, x, \varepsilon)$ (assume $p, H$ at least of class $C^1$ ).

Then $y^2(s) - f(s, x(s), \varepsilon) = y^2(t) - f(t, x(t), \varepsilon) - \int_t^s \frac{\partial f}{\partial t}(\tau, x(\tau), \varepsilon)\, d\tau$ .

Now fix $\varepsilon \sim 0$ to observe the asymptotic behaviour of the eventual solutions of $(P_\varepsilon)$.

3) LEMMA 1. <u>If</u> $x(\tau)$ <u>is finite on some interval</u> $[t, s]$ <u>with</u> $t \sim s$ , <u>then</u> $y^2(s) - f(s, x(s), 0) \sim y^2(t) - f(t, x(t), 0)$ .

<u>Proof</u>. As $\frac{\partial f}{\partial t}$ is standard continuous, $\frac{\partial f}{\partial t}(\tau, x(\tau), \varepsilon)$ is finite on $[t, s]$ ;

hence $\int_t^s \frac{\partial f}{\partial t}(\tau, x(\tau), \varepsilon)\, d\tau \sim 0$, which ends the proof since $f(\tau, x(\tau), \varepsilon) \sim f(\tau, x(\tau), 0)$.

Now assume $p$ bounded (say $p \leq 1$ as in $(F_2)$). Then $\varepsilon p(t, x(t), \varepsilon) \sim 0$ and on any time interval where $y(t)$ and $\dot{y}(t)$ are finite, we have $y(t) \sim 0$ and $H(t, x(t), \varepsilon) \sim H(t, {}^{\circ}(x(t), 0) \sim 0$. As $H$ is standard, the shadow $g = {}^{\circ}x$ is a solution of $H(t, g(t), 0) = 0$. This proves

LEMMA 2. <u>Any slow motion in the finite</u> $(t, x, y)$-<u>space occurs in the halo of some (standard) curve</u> $(t, g(t), 0)$ <u>such that</u> $H(t, g(t), 0) = 0$.

Thus any finite integral curve goes along some equipotential $y^2 = f(\alpha, x, 0)$ + constant as long as $x$ is not in the halo of some singular point of $f(\alpha, \cdot, 0)$ (i.e. a zero of $H(\alpha, \cdot, 0)$ ); then it may loose time along some curve $(t, g(t), 0)$ as in lemma 2 and again start to jump near some time $\beta$ (in very degenerate cases we may have $\beta \sim \alpha$ ).

This rough description shows that a finite solution of $(P_\varepsilon)$ (which must loose one time unit) moves entirely in the halo of the standard surface of equation $y^2 = f(t, x, 0) - f(t, g(t), 0)$, where $g$ is a solution of the reduced equation. <u>Hence any transition between two such solutions</u> $g_1$ <u>and</u> $g_2$ <u>runs along some arc in the intersection of both corresponding surfaces.</u>

<u>Thus all possible asymptotic behaviours may be described in terms of the geometry of these surfaces.</u>

In particular, conditions $F_1$ to $F_6$ are nothing else than geometric restrictions on such surfaces. We shall now discuss a particular situation where the existence and asymptotic behaviour of some solutions is a consequence of qualitative observation. Further we deduce Fife's theorem by means of a geometric device.

3) <u>The bridge bifurcation</u>. Assume that $H(t, x, 0)$ has three standard solutions $g_1(t), g_2(t), c(t)$ on $[0, 1]$, which are continuous and satisfy $g_1(t) < c(t) < g_2(t)$. No other solutions are allowed, for any time $t$.

Put $f_i(t, x, \varepsilon) = \int_{g_i(t)}^{x} 2H(t, \xi, \varepsilon)\, d\xi$ for $i = 1, 2$ and assume that for each $t$, $g_1(t)$ and $g_2(t)$ are minima of $f_i(t, x, 0)$ ; then $c(t)$ is clearly a maximum.

Call $S_i$ the surface of equation $y^2 = f_i(t, x, 0)$ and $S_i(t)$ its section by the $(x, y)$-plane through $(t, 0, 0)$.

Assume that $S_1(t) \cap S_2(t) \neq \emptyset$ for $t \neq t_o$ and $S_1(t_o) = S_2(t_o)$. In other words

$$f_1(t, g_2(t), 0) = -f_2(t, g_1(t), 0) = \int_{g_1(t)}^{g_2(t)} 2H(t_o, x, 0)\, dx \text{ is } \begin{cases} \neq 0 & \text{if } t \neq t_o \\ = 0 & \text{if } t = t_o \end{cases}.$$

Thus we have two cases, according to the relative positions of the curves $S_i(t)$ on both sides of $t_o$. In the first case $f_1(t, g_2(t), 0)$ is $>0$ for $t < t_o$ and $<0$ for $t > t_o$, and we get the bifurcation described in fig. 1, whose critical step is the "bridge" at time $t_o$.

In the second case, $f_1(t, g_2(t), 0)$ is $<0$ before $t_o$ and $>0$ after it. But if we reverse the time around $t_o$, the global picture (and the boundary value problem) is the same, provided 0 and 1 (and also $g_1$ and $g_2$) are interchanged. Hence we may restrict our discussion to the first case and follow the arguments on the fascinating fig. 2.

Call $\bar{a}$ the unique solution of $f_2(0, \bar{a}, 0) = 0$, $\bar{a} > g_1(0)$ and $\bar{b}$ the unique solution of $f_2(1, \bar{b}, 0) = 0$, $\bar{b} < g_2(1)$. The discussion strongly depends on the relative positions of $a, \bar{a}$ and $b, \bar{b}$.

4) The forced case $a < \bar{a}$, $b > \bar{b}$.

THEOREM 1. If $a < \bar{a}$ and $b > \bar{b}$, problem $(P_\varepsilon)$ has a solution and any solution is finite with boundary layers at 0 and 1 along $S_1$ (resp. $S_2$), slow arcs along $(t, g_1(t))$ on $(0, t_o)$ and $(t, g_2(t))$ on $(t_o, 1)$. The layers have thickeness order $\varepsilon$.

Thus there is an unique geometric shadow for the solutions (clearly if $a = g_1(0)$ or $b = g_2(1)$, the corresponding layer fails) as in fig. 3.

The classical translation is clear : there is a family $x_\varepsilon(t)$ with the behaviour expected in Fife's theorem.

Proof. - existence : consider the integral curve $\gamma(t)$ starting at $(0, a, y_o)$. Assume $a < g_1(0)$ (a similar argument works for $a \geq g_1(0)$ ) and put $k = f_1(0, a, 0)$; we have $k > 0$ by hypothesis.

If $y_o > \sqrt{k}$, not $\sim$, then by lemma 1, the curve $\gamma(t)$ remains in the halo of the

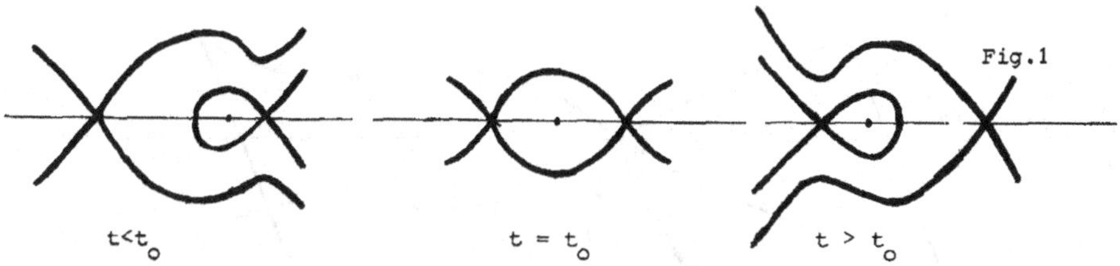


Fig.1

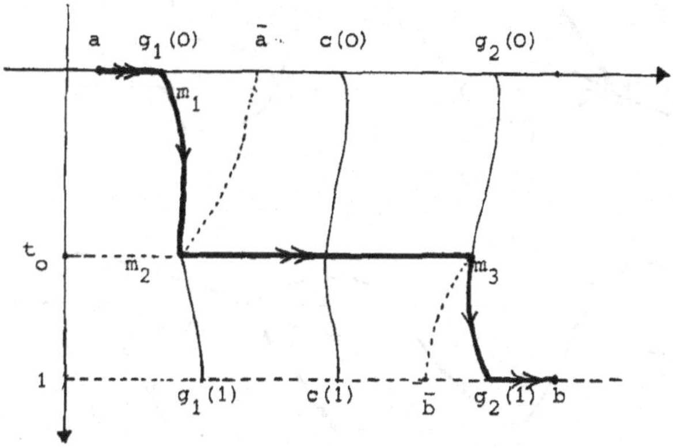


Fig.3

corresponding equipotential of equation $y^2 - f_1(0, x, 0) = y_o^2 - f_1(0, a, 0)$ in the $t=0$ plane, at least as long as $x(t)$ is finite, and the time spent is infinitesimal. These properties are permanent until some infinitely large value of $x(t)$ ; but there $y(t)$ is positive, i.e. $x(1)$ is infinitely large ; hence $x(1) > b$, since $b$ is standard.

If $y_o < \sqrt{k}$, not $\sim$, the same argument yields $x(1)$ infinitely large negative, hence $< b$. By continuity of the flow, there is some $y_o \sim k$ for which $x(1) = b$, i.e. a solution of $(P_\varepsilon)$.

- <u>finiteness</u> : consider a solution $\gamma(t) = (t, x(t), y(t))$ of $(P_\varepsilon)$. If $x(t)$ takes infinitely large values, then $\sup x(t)$ is infinitely large. As $H$ is $>0$ after $\sup x(t)$ is reached, $x(t)$ cannot come back to $x(1) = b$. Similar argument for negative values.

Thus $x(t)$ is finite on $[0, 1]$.

- <u>behaviour</u> : by lemma 1, $\gamma(t)$ starts near $V_a \cap S_1(0)$, i.e. for $a < g_1(t)$ (similar argument for $a \geq g_1(t)$), at $y_o \sim \sqrt{k}$. Using the permanence principle, we see that $\gamma(t)$ jumps along $S_1$ in an infinitesimal time to some point

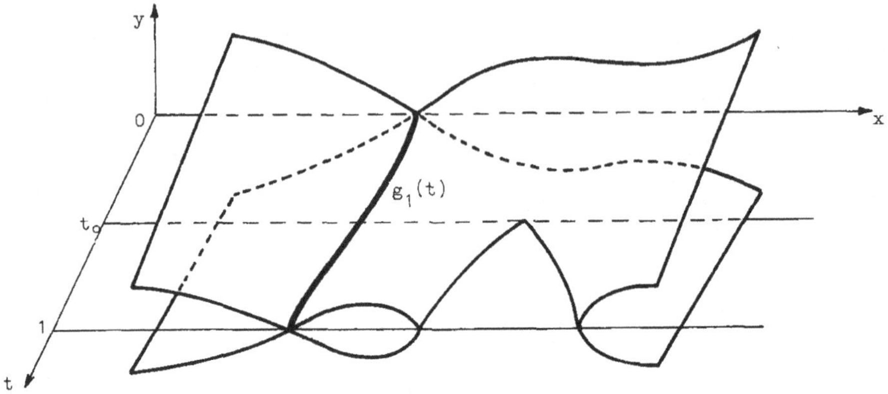


Surface $S_1$.

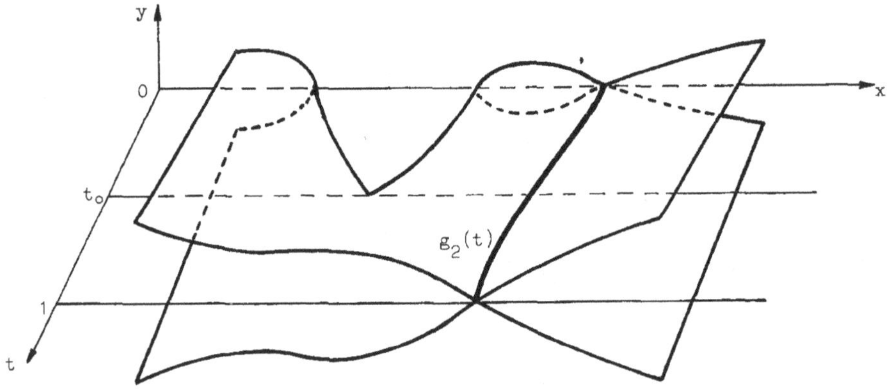


Surface $S_2$.

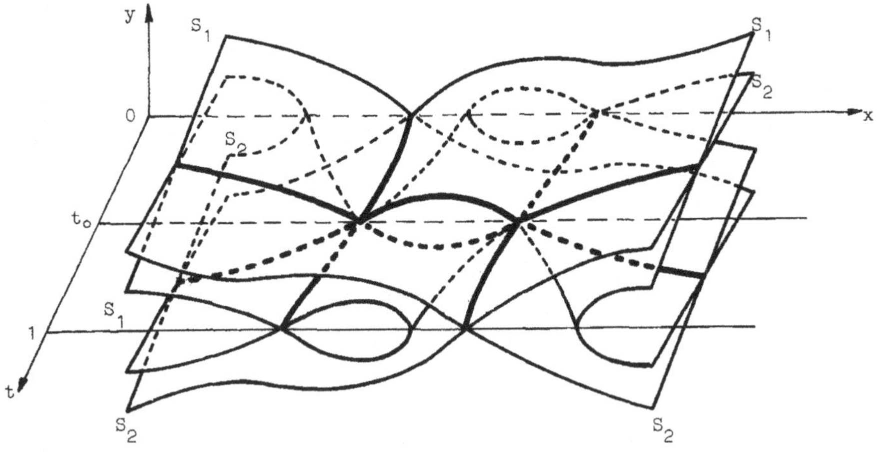


FIG. 2. $S_1 \cup S_2$

$m_1 = (s_1, x_1, y_1) \sim (s_1, g_1(s_1), 0)$ . From there, it cannot leave the halo of $(t, g_1(t), 0)$ before $t \sim t_o$ , since otherwise it would quickly move out of the finite space (and $x(t)$ remains finite) ; thus for any $t < t_o$ , not $\sim$ , we have $(x(t), y(t)) \sim (g_1(t), 0)$ .

Now, for $t > t_o$ , not $\sim$ , it is no longer possible to jump from the line $(t, g_1(t), 0)$ to $V_b$ , since such a jump would occur along $S_2$ , hence never reach $V_b$ due to the hypothesis $b > \overline{b}$ .

We conclude that $\gamma(t)$ must jump along the bridge $S_1(t_o) = S_2(t_o)$ to the line $(t, g_2(t), 0)$ (use again the permanence principle to get precise "corner points" $m_2, m_3$ ; see fig. 3).

Using a similar argument as for $t < t_o$ proves that after $m_3$ , there is a slow motion along $(t, g_2(t), 0)$ until time $s_4 \sim 1$ , where a terminal jump along $S_2$ leads to a point on $V_b$ .

Clearly all layers are of thickeness order $\varepsilon$ , the speed along the jumps being of this order.

All this is now routine for the reader and certainly easier to explain on the blackboard than by means of a written text...

5) From theorem 1, it is easy to deduce Fife's theorem provided in condition $(F_6)$ , $\int_{g_1(t_o)}^{g_2(t_o)} \frac{\partial H}{\partial t}(t_o, x, 0)\, dx < 0$ .

Indeed, conditions $(F_1)$ to $(F_6)$ describe $f_1$ and $f_2$ in some standard neighbourhood N of the expected shadow. An eventual solution of $(P_\varepsilon)$ which moves in N only depends on the values of p and H in N . Thus we may change H outside of N without alterning such a solution.

Now it is clear that Fife's conditions are strong enough (in fact much too strong) to allow a modification of H with the same "germ" along N in order to satisfy all assumptions of § 3 and theorem 1 (for instance $g_2$ may be pushed to the right near $t = 0$ and $g_1$ to the left as to satisfy $a < \overline{a}$ , $b > \overline{b}$ ).

In particular, condition $(F_6)$ (with the negative sign) means that $S_1$ and $S_2$ cross transversaly at time $t_o$ , $S_1$ going under $S_2$ as t increases.

6) If the integral in $(F_6)$ is $>0$, we get a solution of $(P_\varepsilon)$ with inner layer at $t_o$ by time reversing, as explained in § 3. But it first goes along $g_2$ and then along $g_1$, which is not what we want ! To get the expected behaviour in this case, we have to modify theorem 1 (the final geometric device is the same) in order to get solutions with $a > \bar{a}$ and $b > \bar{b}$. But in this case, a lot of different shadows may occur. We list the most typical ones in fig. 4. The reader will easily translate them from the pictures. Notice that solutions with "buckle-layers" may occur, and also very degenerate solutions with an infinitely large number of buckles, provided $\bar{a} < a < g_2(0)$ and $g_1(1) < b < \bar{b}$ (we give up picturing such solutions, for their shadow is two-dimensional...). Now what is important is summarized in

THEOREM 2. For all prototypes listed in fig. 4, there is an actual solution of $(P_\varepsilon)$ with the given behaviour.

We don't give the proof here, refering to the paper [$LS_2$] for details. However the idea is simple : starting on $V_A$ on both sides of surface $S_2$ (in the lower half-$(t=0)$-plane) we get $x(1)$ or $-x(1)$ infinitely large according to $y_o$, which proves the existence of a solution. Now, moving $y_o$ from one side to the other, we get successively by continuity arguments all cases listed in fig. 4 (by lemma 1 and 2, everything happens along $S_1 \cup S_2$ ).

The "buckle-layers" are not too dangerous unexpected layers ; but in some case their number may be arbitrary, with a location strongly depending on the time dependence of $H$ and $p$.

Summarizing our study, we see that Fife's conditions have a geometric meaning, which allows to deduce the qualitative part of his theorem from the observation of the "bridge-bifurcation".

This observation shows a lot of other solutions, even more complicated than in the autonomous problem, "forced" and "unforced" layers occuring simultaneously (but all unforced layers are buckles, due to the uniqueness of the bridge).

Thus the morals is safe : the non autonomous global problem is more difficult than the autonomous one.

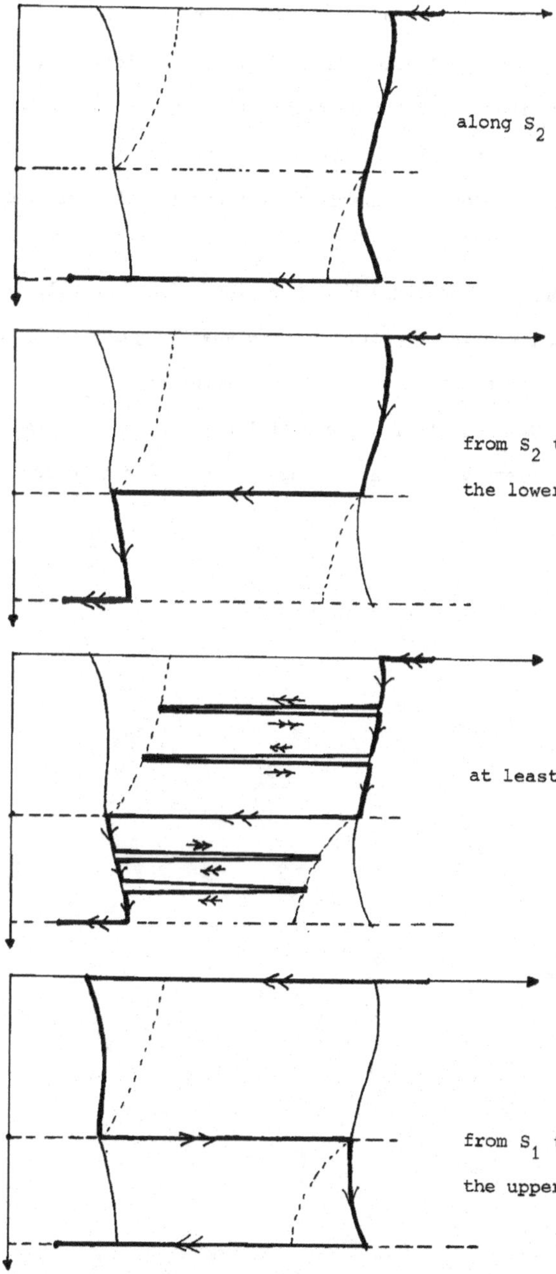

along $S_2$

from $S_2$ to $S_1$ through the lower-bridge.

at least one buckle-jump

from $S_1$ to $S_2$ through the upper-bridge

Fig.4

Just a last remark, which is close to one of G.F. CARRIER in "Singular perturbations theory and geophysics", in Studies in Applied Math. M.A.A., Prentice-Hall, Inc.
A purely existential technic like asymptotic developments cannot describe the whole behaviour of all solutions in a particular problem (i.e. with p and H of some particular type). In the present problem, this is very typical : the only behaviour which is common to all cases is Fife's one.
We hope that the geometric approach outlined here, by means of N.S.A. will lead to some advancement. In any case, drawing a picture to see what is going on is a pleasure that should no longer be despised in the study of asymptotics.
Of course, this is not so easy for partial differential problems as for ordinary ones. But any hope is not completely lost, as we suggest in our next lesson

Lesson 15

IRONING IN THE PROBLEM $\varepsilon^2 \Delta \varphi = \varphi$

<u>Problem.</u> Describe the asymptotic behaviour of the eventual solutions in problem

$$(P_\varepsilon) \qquad \begin{cases} \varepsilon^2 \Delta \varphi = \varphi \\ \varphi|_{\partial G} = h \text{ prescribed function} \end{cases}$$

<u>as</u> $\varepsilon$ <u>tends to</u> 0, <u>where</u> G <u>is some compact domain of</u> $\mathbb{R}^n$ <u>and</u> $\Delta$ <u>the Laplace</u> <u>operator</u> $\Sigma \dfrac{\partial^2}{\partial x_i^2}$ .

THEOREM. <u>Assume that</u> $P_\varepsilon$ <u>has a solution</u> $\varphi_\varepsilon$ <u>of class</u> $c^2$ <u>for</u> $\varepsilon$ <u>little enough.</u> <u>Then this solution is unique and</u>, <u>as</u> $\varepsilon$ <u>tends to</u> 0 , <u>it has the following beha-</u> <u>viour</u> :

       i) <u>on every closed subset</u> $K \subset G - \partial G$ , <u>and for any</u> p , $\varphi_\varepsilon(x)$ <u>is uni-</u> <u>formly of order</u> $\varepsilon^p$ , <u>i.e.</u> $|\varphi_\varepsilon(x)| < \varepsilon^p \alpha(K)$ <u>for every</u> $x \in K$ .

       ii) <u>at every nice point of</u> $\partial G$ , <u>the normal layer thickeness is at most</u> <u>of order</u> $\varepsilon$ .

<u>Comments.</u> 0) Here we make a little attempt in the difficult subject of singular perturbations in partial differential equations. Whatever type you have, elliptic, parabolic or hyperbolic, linear or not, you always work hard in order to get some information. Just look at some papers like R.E. O'MALLEY Jr. : Topics in singular Perturbations. Advances in Math., 2 (1968), p. 365–470.

W. ECKHAUS – E.M. de JAGER : Asymptotic solutions of singular perturbation problems for linear differential equations of elliptic type, Arch. Rat. Mech. Anal. 23 (1966), p. 26–86.

or P.C. FIFE – W.M. GREENLEE : Interior transition layers for elliptic boundary va- lue problems with a small parameter. Russian Math. Surveys, 29 ; 4 (1974), p. 103– 131.

among others to get an insight on such problems.

Of course, non standard tricks are not so easy to find for P.D.E., as for ordinary differential equations, because the underlaying geometry is not obvious : there is nothing like flows of vector fields with nice existence theorems. Moreover an information like " $\Delta\varphi$ is infinitely large" is not as powerful as in the ordinary case, as regards the shape of the domains on which this happens : a "little time interval" is more precise than a "little volume in $\mathbb{R}^n$ " (but if we want only informations "in norm" for convergence of solutions, it may be useful).

For each type of equation, we must find some geometric idea that may replace on usual arguments. Fortunately, there is one in the present case and maybe it could help in some other problems.

The classical treatment of the problem above is not very difficult as long as $G$ is a convex and bounded by a regular hypersurface in $\mathbb{R}^n$ ; in this case, the existence of $\varphi_\varepsilon$ is a consequence of general theorems on linear elliptic equations, provided $h$ is smooth enough. You compute an inner expansion near $\partial G$ using normal coordinates and match it together with an outer expansion of type $\varepsilon^2 v_\varepsilon$ to get a formal solution $\overline{\varphi}_\varepsilon$ ; then you apply the maximum principle to estimate $\varphi_\varepsilon - \overline{\varphi}_\varepsilon$ . Whenever $\partial G$ is not regular (e.g. if there are corners or other pathological points on it), the work becomes very hard, and if, moreover, $G$ is not convex, the trouble may increase.

Here we assume the existence of a solution, without other conditions on $h$ and $G$ :

1) Thus consider a compact connected subset $G$ of $\mathbb{R}^n$ , and assume that on $G$ some function $\varphi_\varepsilon$ of class $C^2$ satisfies $\varepsilon^2\Delta\varphi_\varepsilon = \varphi_\varepsilon$ , $\varphi_\varepsilon|_{\partial G} = h$ (here $\partial G = G - \overset{\circ}{G}$ may be quite complicated).

Due to the maximum principle, $\varphi_\varepsilon$ <u>is unique</u>.

Indeed, if $\varphi_\varepsilon + \psi$ is another solution, we have $\varepsilon^2\Delta\psi = \psi$ and $\psi|_{\partial G} = 0$ . If $\psi$ takes positive values, it takes its maximum at an interior point $a$ , and all their derivatives $\frac{\partial^2\psi}{\partial x_i^2}$ are $\leq 0$ ; hence $\psi(a)$ would be $\leq 0$ , which is not possible.

The same argument works for negative values.

We have to study $\varphi_\varepsilon$ as $\varepsilon \to 0$ .

2) Assertion (i) implies that the family $\varphi_\varepsilon$ has boundary layer character in a sense which clearly generalizes that of lesson IV.10.
As for thickeness, some care is needed if $\partial G$ is not a regular hypersurface. "Nice points" will be defined further, in relation with our geometric approach.
First transfer the problem, i.e. assume that $G$, $h$ (hence $\partial G$), and $n$ are standard and take $\varepsilon \sim 0$.
Put $m = \inf h$ and $M = \sup h$ on $\partial G$ and write $\varphi$ for $\varphi_\varepsilon$.
Consider a closed euclidian ball $B(a, 2r)$ with center $a$ (standard or not) and standard radius $2r > 0$ such that $\mathring{B} \subset \mathring{G}$ (see fig. 1).

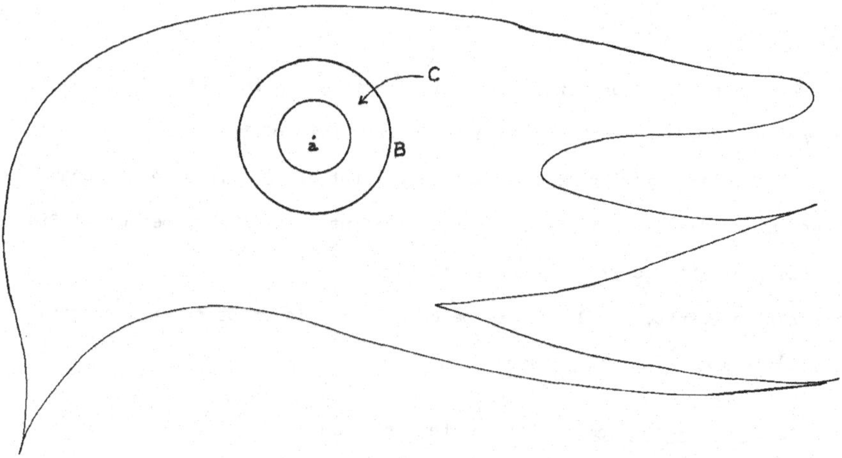


Call $C(a, r)$ the closed spherical crown $B(a, 2r) - \mathring{B}(a, r)$ : For any $x \in C$, put $d(x, \partial C) = \inf(2r - \|x - a\|, \|x - a\| - r)$. Our main tool is the

3) Ironing lemma.

i) For every $x \in \mathring{C}$ such that $d(x, \partial C)$ is not $\sim 0$, $\frac{1}{\varepsilon^p}\varphi(x)$ is finite for any standard $p$.

ii) For every $x \in \mathring{C}$ such that $\frac{1}{\varepsilon} d(x, \partial C)$ is infinitely large, one has $\varphi(x) \sim 0$.

In other words, the restriction of $\varphi$ to $C$ has boundary layer character of thickeness order $\varepsilon$ in the normal direction along $\partial C$.

Before proving this lemma, let us deduce the theorem from it.

The principle is clear : take your flat "iron" $C(a, r)$ and move it within $G$ ; if necessary, change the radius (but it must remain standard) to iron as far as possible (recall that $\overset{\circ}{C}$ must be in $\overset{\circ}{G}$ ).

Then if ${}^{\circ}x \in \overset{\circ}{G}$ , there is a crown $C(a, r)$ of the wanted type such that $d(x, \partial C)$ is not $\sim 0$ (use a standard ball in $\overset{\circ}{G}$ containing ${}^{\circ}x$ ). Hence by i), $\frac{1}{\varepsilon^2}\varphi(x)$ is finite.

Now any point in a standard closed subset $K \subset \overset{\circ}{G}$ satisfies ${}^{\circ}x \in \overset{\circ}{G}$ , for ${}^{\circ}x \in K$ .

By transfer, this proves part (i) of the theorem, the $\sup\limits_{\text{on } K}$ of $\frac{\varphi(x)}{\varepsilon^P}$ is finite (if reached somewhere or not) ;

thus for all $\varepsilon$ , $\sup \frac{\varphi(x)}{\varepsilon^P}$ is standard.

For a precise formulation of part (ii), call "nice" all points of $\partial C$ which may be reached by our ironing technic and then apply part (ii) of the lemma.

If $\partial G$ is a regular hypersurface, clearly any point of $\partial G$ is nice. A corner (or some point on an edge) is not nice, nor is any point in its halo, because infinitesimal radius would be necessary to reach them.

Thus at non-nice points, the layer may be of highęr thickeness order (but still infinitesimal) in some transversal directions.

4) <u>Proof of the lemma</u>. Let $\bar{\varphi}$ be the eventual solution on $C$ of $\begin{cases} \varepsilon^2 \Delta\bar{\varphi} = \bar{\varphi} \\ \bar{\varphi} = M \text{ on } \partial C \end{cases}$ . Then $\psi = \bar{\varphi} - \varphi$ is $\geq 0$ on $\partial C$ , for $C \subset G$ . Either $\inf \psi$ is taken at a boundary point, hence is $\geq 0$ , or at an interior point and again it is $\geq 0$ (maximum principle). Thus, we have $\varphi \leq \bar{\varphi}$ . Similarly $\bar{\bar{\varphi}} \leq \varphi$ , where $\bar{\bar{\varphi}}$ is the eventual solution of $\begin{cases} \varepsilon^2 \Delta\bar{\bar{\varphi}} = \bar{\bar{\varphi}} \\ \bar{\bar{\varphi}} = m \text{ on } \partial C \end{cases}$ .

If we prove the existence of $\bar{\varphi}$ and $\bar{\bar{\varphi}}$ with the behaviour described in the lemma, we infer properties (i) and (ii) for $\varphi$ .

But both equations have rotationnal symmetry, that is $\bar{\varphi}$ and $\bar{\bar{\varphi}}$ are certainly functions of the radial variable only.

Put $a = (a_1, \ldots, a_n)$ and $t = \sqrt{(x_1 - a_1)^2 + \ldots + (x_n - a_n)^2}$ , which is a $C^\infty$ function on $C(a, r)$ . Consider $\overline{\varphi}(x) = f(t(x))$ . We have $\Delta\overline{\varphi} = f'' + \frac{n-1}{t} f'$ and fortunately, we get the ordinary problem

$$\begin{cases} \varepsilon^2(f'' + \frac{n-1}{t} f') = f \quad \text{on} \quad [r, 2r] \\ f(r) = f(2r) = M , \end{cases}$$

that we may study as in former lessons. In the stretched phase space

$$\begin{cases} f' = \frac{y}{\varepsilon} \\ y' = \frac{f}{\varepsilon} - \frac{k}{t} y \quad \text{with} \quad k, r, M \quad \text{standard.} \\ t' = 1 \\ x(r) = x(2r) = M . \end{cases}$$

We have to prove that

i) for any $t \in ]r, 2r[$ , such that $t$ not $\sim r$ and not $\sim 2r$ , and any standard $p$ , $\frac{1}{\varepsilon^p} f(t)$ is finite.

ii) for any $t$ such that $\frac{t-r}{\varepsilon}$ and $\frac{2r-t}{\varepsilon}$ are both infinitely large, $f(t)$ is infinitesimal.

We have $ff' - yy' = \frac{k}{t} y^2$ ; hence the integral curve $(t, f(t), y(t))$ starting at $(t_0, f_0, y_0)$ satisfies $f^2(t) - y^2(t) \sim f_0^2 - y_0^2$ as long as $y(t)$ is finite and $t \sim t_0$ . This is nearly a prime integral for "fast arcs". Start at $(r, M, y_1)$ with $y_1$ standard, $-M < y_1$; then $f(2r)$ is infinitely large positive. Start at $(r, M, y_2)$ with $y_2$ standard, $y_2 < -M$ ; then $f(2r)$ is infinitely large negative (see fig. 2). Due to the continuity of the flow with respect to $y_0$ , there exists an $y_0 \sim M$ for which $f(2r) = M$ , i.e. a solution for the boundary value problem.

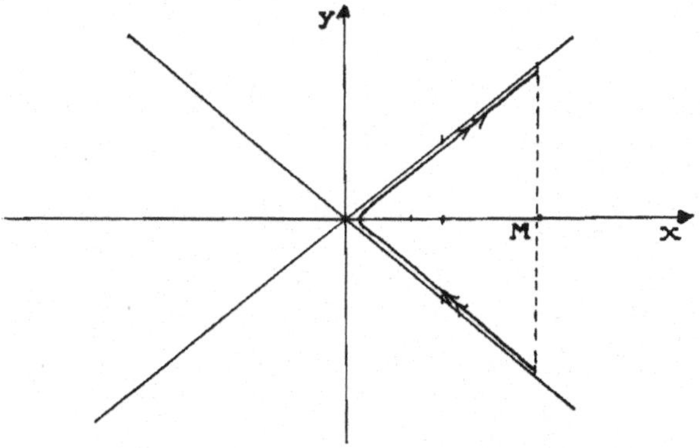

Consider such a solution. Suppose that at some $t_o \in ]r, 2r[$ , not infinitely close to $r$ or $2r$ , $\frac{f(t_o)}{\varepsilon}$ is infinitely large ; if $y(t_o) \geq 0$ , then $\frac{f(t)}{\varepsilon}$ remains infinitely large for $t > t_o$ and hence the time spent between $f(t_o)$ and $M$ is infinitesimal, which is not the case ; if $y(t_o) < 0$ , the same arguments proves that $t_o \sim r$ . Thus for any $t$ not $\sim r$ or $\sim 2r$ , $\frac{f(t)}{\varepsilon}$ is finite.

Now consider $t > s$ , $t$ not $\sim r$ , $s$ not $\sim 2r$ ; for any standard $p$ , we may find standard points $t_i , s_i$ $(i = 1 , \ldots , p)$ such that

$$r < t_1 < \ldots < t_p < t < s < s_p < \ldots < s_1 < 2r .$$

Call $f_i = \frac{f}{\varepsilon^i}$ ; $f_i$ is a solution on $[t_i , s_i]$ of the equation $\varepsilon^2(\ddot{f}_i + \frac{k}{t}\dot{f}_i) = f_i$ .

The argument above proves that if $f_i(t_i)$ and $f_i(s_i)$ are finite, so is $f_{i+1}(t)$ on $[t_{i+1} , s_{i+1}]$ .

Thus, by recurrence, we get $\frac{f}{\varepsilon^p}$ finite on $[t , s]$ .

As for (ii), just remark that the speed of the solution is of order $\frac{1}{\varepsilon}$ as long as $f(t)$ is not $\sim 0$ . Hence for $\frac{t-r}{\varepsilon}$ and $\frac{2r-t}{\varepsilon}$ infinitely large, $f(t)$ is infinitesimal.

## GENERAL REFERENCES ON NON STANDARD ANALYSIS


DAVIS M. Applied non standard analysis , New-York , Wiley , 1977 .

FENSTAD J.E. Non standard méthods in Stochastic Analysis ant Mathématical Physics , Iber.d.Dt.Math.Verein., 82 ,(1980) , 167-180.

HURD A. , LOEB P. Victoria Symposium on Non Standard Analysis . Springer Lecture Notes , 1974 , n° 369 .

KEISLER H. J. Foundations of infinitesimal calculus , Boston , Prindle , Weber et Schmidt , 1976 .

KREISEL G. An application of model theory to algebra , analysis , probability . Inter.Sympos.Pasadena , Californie 1967 , Holt , Rinehart and Winston , New-York , 1969 .

LAUGWITZ D. Infinitesimalkalkül , Mannheim : B.I. Wissenschaftsverlag 1978 .

LAUGWITZ D. The theory of Infinitesimals ; an introduction to N.S.A. Roma 1980 , Acad.Naz.dei Lincei .

LUXEMBURG W.A.J. A general theory of monads . Application of model theory to algebra , analysis , probability ( Inter.Sympos.Pasadena , Californie 1967 Holt , Rinehart and Winston , New-York (1969) .

LUXEMBURG W.A.J. - ROBINSON A. Contributions to Non Standard Analysis , North-Holland Publisching Comp., Amsterdam-London , 1972 .

NELSON E. Internal Set Theory : a new approach to non standard analysis . Bull.Amer. Math.Soc , 83 (1977) 1165-1198 .

ROBINSON A. Non standard analysis , Amsterdam , North-Holland , 1966 .

ROBINSON A. Collected Works , Vol.2 , Amsterdam , North-Holland 1979 .

SCHMIEDEN C. - LAUGWITZ D. Eine Erweiterung der Infinitesimalrechnung , Math. Zeitschr. 69 (1959) , p.1.

STROYAN K.D. - LUXEMBURG W.A.J. Introduction to the theory of infinitesimals . New-York-London , Academic Press 1976 .

TECHNICAL REFERENCES RELATED WITH SECTION IV


BEBBOUCHI R. Equations différentielles ordinaires . Propriétés topologiques de l'ensemble des solutions passant par un point . I.R.M.A. (1980)

BENOIT E. Equation de Van der Pol avec terme forçant . I.R.M.A. (1979)

BENOIT E. Tunnels et entonnoirs . C.R.Acad.Sc.Paris , to appear .

BENOIT E. , CALLOT J.L. , DIENER F. , DIENER M. Chasse au canard . I.R.M.A. (1980)

BOBO S. Ombre d'un polynôme de degré standard . I.R.M.A. (1979)

CALLOT J.L. Géodésiques des cubes , polyèdres , billards vus sous l'angle de l'Analyse Non Standard . I.R.M.A. (1978)

CALLOT J.L. Un point de vue non standard sur l'équation d'Hermite $\ddot{X} - T\dot{X} + nX = 0$ I.R.M.A. (1981)

CALLOT J.L , DIENER F. , DIENER M. Le problème de la"chasse au canard" C.R.Acad.Sc. Paris , t.286 , Serie A , 1059-1061 .

DIENER F. Famille d'équations à cycle limite unique , C.R.Acad.Sc.Paris ,289, serie A (1979) , 571-574 .

DIENER F. Les équations $\varepsilon\ddot{x} + (x^2-1)\dot{x}^{[s]} + x = a$ Collectanea Mathematica , vol XXIX , fasc.3 (1978)

DIENER F. Quelques exemples de bifurcations et leurs canards . I.R.M.A. (1979)

DIENER F. Les canards de l'équation $\ddot{y} + (\dot{y} + a)^2 + y = 0$ I.R.M.A. (1980)

DIENER M. Perturbations singulières des systèmes de Liénard . I.R.M.A. (1978)

DIENER M. Mais qu'est-ce donc que des canards? I.R.M.A. (1979)

DIENER M. Nessie et les canards I.R.M.A. (1979)

DIENER M. Deux nouveaux phènomènes canard . C.R.Acad.Sc.Paris , 290 , série A (1980)

DIENER M. , VAN DEN BERG I. Halos et galaxies . To appear C.R.Acad.Sc.Paris .

GOZE M. Algèbres de Lie modèles et déformation . To appear C.R.Acad.Sc.Paris .

GOZE M. Modèles d'algèbres de Lie Symplectiques . To appear I.R.M.A.

GOZE.M, LUTZ R. Pratique commentée de la méthode non classique . I.R.M.A. (1980)

HARTHONG J. Le Moiré . 2ème édition , I.R.M.A. (1980)

HARTHONG J. Les singularités des fonctions spéciales sur une variété riemannienne infiniment aplatie . Séminaire Goulaouic-Schwarz .

HARTHONG J. Formule de Poisson pour les variétés à bord ; une nouvelle méthode inspiréé par G.D.Birkhoff . I.R.M.A. (1979)

LUTZ R., SARI T. Sur le comportement asymptotique des solutions dans un problème non linéaire . To appear .

LUTZ R., SARI T. Sur le comportement asymptotique des solutions dans un problème aux limites non autonomes . To appear .

REEB G. Séance -débat sur l'Analyse non Standard . Gazette des Mathématiciens n°8 (1977) p8-14 .

REEB G. La Mathematique Non Standard vieille de soixante ans ? I.R.M.A. (1979)

REEB G. Formalisme , intuitionnisme , analyse non standard et diverses situations paradoxales qui y sont liées . I.R.M.A. (1979) .

REEB G. Equations différentielles et analyse non classique ( d'après J.L.Callot) , Proceeding of the IV international colloquium of differential geometry , Santiago de Compostella , 1978 , p.240-245.

REEB G. Un principe de Transfert en mathématiques classiques . I.R.M.A. (1980)

REEB G., SCHWEITZER P. Un lemme de Thurston établi au moyen de l'analyse Non Standard Rio de Janeiro , 1977 , Springer Lecture Notes n° 652

REEB G., TROESCH A. URLACHER E. Analyse non Standard . I.R.M.A. (1978) Seminaire LOIS

SARI T. Sur le comportement asymptotique des solutions dans un problème quasi-linéaire . To appear .

TROESCH A. Thèse Strasbourg 1981.

TROESCH A., URLACHER E. Analyse non Standard et l'équation de Van der Pol .I.R.M.A. 1977 .

TROESCH A., URLACHER E. Perturbations singulières et analyse non standard . $C^k$ convergence et crépitement des solutions . I.R.M.A. (1977)

TROESCH A., URLACHER E. Analyse non Standard Et l'equation de Van der Pol . C.R.Acad Sc.Paris , t.286 , 1109-1111 and t.287 , 937-939 and I.R.M.A. (1978)

URLACHER E. Equations Différentielles du type $\varepsilon\ddot{x} + f(\dot{x}) + x = 0$ , avec $\varepsilon$ petit ; I.R.M.A. (1980) .

---

I.R.M.A. Papers I.R.M.A. are available and can be requested at

I.R.M.A.

10 , rue du Général Zimmer

67084 STRASBOURG CEDEX

FRANCE .

# AUTHOR INDEX

## GLOSSARY

Accumulation point I.6.
Approachable point III.4.
Artin Schreier theorem II.5.4.
Assemblage II.1.3.
Atomic formula II.1.3.
Axiom of choice II.3.3., II.6
Algebraic topology III.10
Approximatively linear I.52.
Ascoli's lemma III.5.3.
Asymptotic behaviour IV.10
Axiom ... II.1.6.

Bernstein polynomial III.6.4.
Bolzano-Weierstrass theorem I.6., III.3.2.
Boundary-value problem IV.10
Bracket of vector fields III.7.5.
Big halo III.1.2.
Boundary layer IV.10.0.
Boundedness of integral curves IV.6.
Brouwer's theorem III.6.3.

Canards IV.8.
Characteristic vector IV.4.5.
Compactness III.3
Concurrent II.4.3.
Construction principle II.7.7.
Continuity I.2., III.1.5.
Continuous (S) I.4., III.4.8
Contraction of Lie algebras IV.4.
Convergent (*) I.10
Cardinal number II.6.4.
Cohomology III.10
Complete (field) I.11
Consistent II.1.9, II.6.4.
Contact form IV.4.5.
Continuous (neo) I.5.2., III.4.8
Continuous (*) I.R., III.1.6, III.4.8.
Convergent (S) I.10.

Deformation of Lie algebras I.V.4
Differentiability I.5.
Degree of solution IV.13.3.
Differentaible (function) I.4

Enlargements I.1.5, I.1.art.1, II.4, II.6
Extension I.1. art. 1
Extension (conservative) II.1.7.
Extension of first order structure II.4.2.
External set III.2
Equicontinuous III.5
Extension (by definition) II.1.7.
Extension non-standard) II.3.
External formula III.7.5.